COGNITIVE WORK ANALYSIS:
COPING WITH COMPLEXITY

Human Factors in Defence

Series Editors:

Dr Don Harris, Cranfield University, UK
Professor Neville Stanton, Brunel University, UK
Professor Eduardo Salas, University of Central Florida, USA

Human factors is key to enabling today's armed forces to implement their vision to 'produce battle-winning people and equipment that are fit for the challenge of today, ready for the tasks of tomorrow and capable of building for the future' (source: UK MoD). Modern armed forces fulfil a wider variety of roles than ever before. In addition to defending sovereign territory and prosecuting armed conflicts, military personnel are engaged in homeland defence and in undertaking peacekeeping operations and delivering humanitarian aid right across the world. This requires top class personnel, trained to the highest standards in the use of first class equipment. The military has long recognised that good human factors is essential if these aims are to be achieved.

The defence sector is far and away the largest employer of human factors personnel across the globe and is the largest funder of basic and applied research. Much of this research is applicable to a wide audience, not just the military; this series aims to give readers access to some of this high quality work.

Ashgate's *Human Factors in Defence* series comprises of specially commissioned books from internationally recognised experts in the field. They provide in-depth, authoritative accounts of key human factors issues being addressed by the defence industry across the world.

Cognitive Work Analysis:
Coping with Complexity

DANIEL P. JENKINS
NEVILLE A. STANTON
PAUL M. SALMON
GUY H. WALKER
Human Factors Intergration Defence Technology Centre

CRC Press
Taylor & Francis Group
Boca Raton London New York

CRC Press is an imprint of the
Taylor & Francis Group, an **informa** business

iPod is a registered trademark of Apple Inc.

CRC Press
Taylor & Francis Group
6000 Broken Sound Parkway NW, Suite 300
Boca Raton, FL 33487-2742

First issued in paperback 2017

© 2009 by Daniel P. Jenkins, Neville A. Stanton, Paul M. Salmon, Guy H. Walker
CRC Press is an imprint of Taylor & Francis Group, an Informa business

No claim to original U.S. Government works

Version Date: 20160226

ISBN 13: 978-0-7546-7026-1 (hbk)
ISBN 13: 978-1-138-07338-8 (pbk)

Visit the Taylor & Francis Web site at
http://www.taylorandfrancis.com

and the CRC Press Web site at
http://www.crcpress.com

Contents

List of Figures

List of Tables

Acknowledgements

The Human Factors Integration Defence Technology Centre is a consortium of defence companies and Universities working in cooperation on a series of defence related projects. The consortium is led by Aerosystems International and comprises Birmingham University, Brunel University, Cranfield University, Lockheed Martin, MBDA, and SEA. The consortium was recently awarded The Ergonomics Society President's Medal for work that has made a significant contribution to original research, the development of methodology, and application of knowledge within the field of ergonomics.

Aerosystems International	Birmingham University	Brunel University
Dr Karen Lane	Professor Chris Baber	Professor Neville Stanton
Dr David Morris	Professor Bob Stone	Dr Guy Walker
Linda Wells	Dr Huw Gibson	Dr Daniel Jenkins
Kevin Bessell	Dr Robert Houghton	Dr Paul Salmon
Kelly Maddock-Davies	Richard McMaster	Amardeep Aujla
Nicola Gibb	Dr James Cross	Kirsten Revell
Robin Morrison	Robert Guest	Laura Rafferty

Cranfield University	Lockheed Martin UK	MBDA Missile Systems
Dr Don Harris	Mick Fuchs	Dr Carol Mason
Andy Farmilo	Lucy Mitchell	Grant Hudson
Geoff Hone	Mark Linsell	Chris Vance
Jacob Mulenga	Ben Leonard	Steve Harmer
Ian Whitworth	Rebecca Stewart	David Leahy
John Huddlestone		
Antoinette Caird-Daley		

**Systems Engineering and
Assessment (SEA) Ltd**

Dr Anne Bruseberg

Dr Iya Solodilova-Whiteley

Mel Lowe

Ben Dawson

Georgina Fletcher

We are grateful to DSTL who have managed the work of the consortium, in particular to Geoff Barrett, Bruce Callander, Jen Clemitson, Colin Corbridge, Roland Edwards, Alan Ellis, Jim Squire, Alison Rogers and Debbie Webb.

This work from the Human Factors Integration Defence Technology Centre was part-funded by the Human Sciences Domain of the UK Ministry of Defence Scientific Research Programme.

Further information on the work and people that comprise the HFI DTC can be found on www.hfidtc.com.

Thanks to Dr Mark Young for allowing us the opportunity to present a guest lecture on his undergraduate course, this provided us with the feedback required to develop an understanding of how to communicate and teach the ideas behind Cognitive Work Analysis.

Many thanks to Neelam Naikar of Defence Science and Technology Organisation (DSTO), Australia for the frequent feedback and reassurance particularly on the analysis of the 'sensor to effecter paradigm' presented in Chapter 6.

Thanks to Andy Farmilo from the Defence Academy at Shrivenham who developed the HFI DTC CWA tool with us (discussed in Chapter 9). Also a big thanks to all those who gave feedback on the early versions of the tool, particularly; Stewart Birrell, Ben Elix, Greg Jamieson, John Lee, Gavin Lintern, Neelam Naikar, Bobbie Seppelt and David Zeltzer.

Many thanks are extended to the many subject matter experts who provided us with the invaluable domain knowledge required for completing this book. Special acknowledgement is required for Major Mike Forster for his input in the development of the metaphoric bird-table (Chapter 3); Nick Wharmby MBE, Shaun Wyatt DFC, Jan Ferraro and Sean Dufosee for their assistance in both the data collection and the interpretation of the analysis products in the mission-planning example (Chapter 5); and Geoff Hone for his advice on the Armoured Battlegroup in the Quick Attack (Chapter 9).

Final thanks to all of the experimental participants who gave up their time to take part in the various studies described in this book.

About the Authors

Dr Daniel P. Jenkins
HFI DTC, BIT Lab, School of Engineering and Design
Brunel University, Uxbridge
UK
daniel.jenkins@brunel.ac.uk

Dan Jenkins graduated in 2004, from Brunel University, with an M.Eng (Hons) in Mechanical Engineering and Design, receiving the 'University Prize' for the highest academic achievement in the school. As a sponsored student, Dan finished University with over two years experienced as a Design Engineer in the Automotive Industry. Upon graduation, Dan went to work in Japan for a major car manufacturer, facilitating the necessary design changes to launch a new model in Europe. In 2005, Dan returned to Brunel University taking up the full-time role of Research Fellow in the Ergonomics Research Group, working primarily on the Human Factors Integration Defence Technology Centre (HFI-DTC) project. Dan studied part-time on his PhD in human factors and interaction design – graduating in 2008, receiving the 'Hamilton Prize' for the Best Viva in the School of Engineering and Design. Both academically and within industry Dan has always had a strong focus on human factors, system optimisation and design for inclusion. Dan has authored and co-authored numerous journal paper, conference articles, book chapters and books. Dan and his colleagues on the HFI DTC project were awarded the Ergonomics Society's President's Medal in 2008.

Professor Neville A. Stanton
HFI DTC, BIT Lab, School of Engineering and Design
Brunel University, Uxbridge
UK
neville.stanton@brunel.ac.uk

Professor Stanton holds a Chair in Human Factors and has published over 120 international peer-reviewed journal papers and 12 books on Human Factors and Ergonomics. In 1998, he was awarded the Institution of Electrical Engineers Divisional Premium Award for a co-authored paper on Engineering Psychology and System Safety. The Ergonomics Society awarded him the President's medal in 2008 and the Otto Edholm medal in 2001 for his contribution to basic and applied ergonomics research. The Royal Aeronautical Society awarded him the Hodgson Medal and Bronze Award with colleagues for their work on flight deck safety. Professor Stanton an editor of Ergonomics and on the editorial board of

Theoretical Issues in Ergonomics Science and the International Journal of Human Computer Interaction. Professor Stanton is a Fellow and Chartered Occupational Psychologist registered with The British Psychological Society, and a Fellow of The Ergonomics Society. He has a BSc in Occupational Psychology from Hull University, an MPhil in Applied Psychology from Aston University, and a PhD in Human Factors, also from Aston.

Dr Paul M. Salmon
HFI DTC, BIT Lab, School of Engineering and Design
Brunel University, Uxbridge
UK.
paul.salmon@brunel.ac.uk

Dr Paul Salmon is a Research Fellow within the Human Factors Group at Brunel University and holds a BSc in Sports Science and an MSc in Applied Ergonomics, both from the University of Sunderland, and a PhD in Human Factors from Brunel University. Paul has significant experience in applied human factors research in a number of domains, including the military, civil and general aviation, rail and road transport, and has previously worked on a variety of research projects within these sectors. This has led to Paul gaining expertise in a broad range of areas, including human error, situation awareness, and the application of human factors methods, including human error identification, situation awareness measurement, teamwork assessment, task analysis and cognitive task analysis methods. Paul's current research interests include the areas of situation awareness in command and control, human error and the application of human factors methods in sport. Paul has authored and co-authored various scientific journal articles, conference articles, book chapters and books and was awarded the 2006 Royal Aeronautical Society Hodgson Prize for a co-authored paper in the society's journal. He was also co-recipient of the Ergonomic Society's President's Medal in 2008.

Dr Guy H. Walker
HFI DTC, BIT Lab, School of Engineering and Design
Brunel University, Uxbridge
UK.
guy.walker@brunel.ac.uk

Dr Guy Walker read for a BSc Honours degree in Psychology at Southampton University specialising in engineering psychology, statistics and psychophysics. During his undergraduate studies he also undertook work in auditory perception laboratories at Essex University and the Applied Psychology Unit at Cambridge University. After graduating in 1999 he moved to Brunel University, gaining a PhD in Human Factors in 2002. His research focused on driver performance, situational awareness and the role of feedback in vehicles. Since this time Guy has worked for a human factors consultancy on a project funded by the Rail Safety and

Standards Board, examining driver behaviour in relation to warning systems and alarms fitted in train cabs. Currently Guy works within the DTC HFI consortium at Brunel University, engaged primarily in work on future C4i systems. He is also author of numerous journal articles and book contributions. Along with his colleagues on the HFI DTC project, Guy was awarded the Ergonomic Society's President's Medal in 2008.

Commonly Used Analysis Acronyms and Initialisms

The following is a reference list of analysis acronyms and initialisms used within the document.

ABP	Assumption Based Planning
ADS	Abstraction Decomposition Space
AH	Abstraction Hierarchy
CA	Comprehensive Approach
CAT	Contextual Activity Template
ConTA	Control Task Analysis
CWA	Cognitive Work Analysis
EBO	Effects Based Operations
EID	Ecological Interface Design
GUI	Graphical User Interface
HCI	Human Computer Interaction
HF	Human Factors
HTA	Hierarchal Task Analysis
KBB	Knowledge-Based Behaviour
NCW	Network Centric Warfare
NEC	Network Enabled Capability
RBB	Rule-Based Behaviour
ROM	Rough Order of Magnitude
SBB	Skill-Based Behaviour
SOCA	Social Organisation and Cooperation Analysis
SRK	Skills, Rules, Knowledge
StrA	Strategies Analysis
UCD	User Centred Design
WCA	Worker Competencies Analysis
WDA	Work Domain Analysis

Commonly Used Military Acronyms and Initialisms

The following is a reference list of military acronyms and initialisms used within the document.

2IC	Second in Command
Bde	Brigade
BG	Battlegroup
BAE	Battlespace Area Evaluation
C2	Command and Control
C2DC	Command and Control Development Centre
CAST	Command and Staff Trainer
CoA	Course of Action
DCDC	Development, Concepts and Doctrine Centre
FOO	Forward Observation Officer
FSG	Fire Support Group
FUP	Forming Up Position
HQ	Head Quarters
IO	International Organisation
JDCC	Joint Doctrine and Concepts Centre
JDN	Joint Doctrine Note
LAW	Light Anti-armour Weapon
MFC	Mortar Fire Controller
MPS	Mission Planning System
NCW	Network Centric Warfare
NGO	Non-Government Organisation
NEC	Network Enabled Capability
OC	Operational Commander
OPFOR	Opposition Forces
OPV	Observation Post Vehicle
POW	Prisoners of War
REME	Royal Electrical and Mechanical Engineers
UC2	Ubiquitous Command and Control
UDO	User Defined Overlay

Glossary

The following is a short glossary of some of the key terms used throughout the book.

Complexity

Complexity is a qualitative term used describing how complex a particular system or a domain is. It is important to establish the difference between complexity and something that is complicated. According to Woods (1988), complex systems can be defined by four dimensions; they are high risk, dynamic, uncertain systems containing numerous interconnected parts. The term complex is used, in this book, to describe such systems. The term complicated is reserved for systems that my contain many interconnected parts but are not high risk, dynamic or uncertain (For a more complete discussion, see Chapter 2).

Formative

The term 'formative' is used to communicate a development of understanding. In this book, the term is used to describe systems or processes that support the development of knowledge or understanding. (A more complete description, contrasting formative approaches with normative and descriptive approaches, can be found in Chapter 2).

Prescriptive

The word 'prescriptive' is used throughout this book in the discussion of methodologies and framework. A prescriptive approach is considered one that contains directions, injunctions, rules and laws. Approaches defined by a broader philosophy that contain some level of guidance, but are not tightly defined by fixed process and rules, are termed non-prescriptive in this book.

Constraint

The word 'constraint' is used throughout this book to describe a wide range of limitations including physical, contextual, and societal. In this book, constraints focus on the limits of possibility, identifying states where work is, and is not, possible.

Sociotechnical

The term 'sociotechnical' can be broken down into two composing parts: socio relating to the social subsystems and technical relating to technical subsystems. With this simple definition, it is clear that many systems fall into this category. According to Walker et al (2007), the phrase 'sociotechnical' is ubiquitous in Human Factors literature. As a 'theory', it describes the interrelatedness of socio and technical elements of a work domain and the non-linearities that are created when one or other is optimised. Sociotechnical systems theory in, primarily, concerned with the joint optimisation of people and systems and the notion that, system performance is inextricably linked to the experiences of those at work within it.

Chapter 1
Introduction

Humanity's constant strive to do things better, faster and more efficiently continues to be a fundamental driving force for human development. In the modern world, this can be seen by the enormous number of products and services coming to market every single day. Not only do these new products and services allow us to do things better, faster and more efficiently they bring with them new affordances and capabilities; some of which, that were previously unthought-of. Human beings have an innate ability to adapt to these new opportunities and flourish; this allows us to conceive of new ways of using existing systems and designs, which in turn often leads to the design and development of new products and services. The idea of a 'perfect product' is a fallacy, as soon as a new product is exposed to human beings, new unanticipated uses are developed and improvements conceived of. As we as humans develop we expect, if not demand, our products and services to follow suit.

Figure 1.1 shows Carroll's (1991) 'task-artefact cycle'. One way to enhance the longevity as well as the usefulness of a product or service is to design for this innate curiosity and desire for improvement; this involves developing products and services that encourage and support rather than constrain this natural, formative, human behaviour.

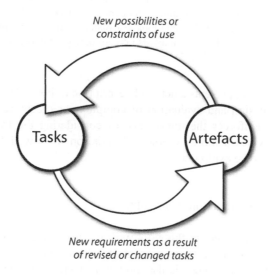

Figure 1.1   The task-artefact cycle (adapted from Carroll, 1991)

This book aims to address how systems can be designed and developed to support, rather then discourage, this kind of formative behaviour. More specifically, this book explores the ability of Cognitive Work Analysis (CWA) to support the analysis, development and evaluation of complex sociotechnical systems.

This book builds heavily upon previous works on CWA and Work Domain Analysis (most notably; Rasmussen et al, 1994; Vicente, 1999; Naikar et al, 2005). Whilst there is concordance with Vicente's (1999) view, that CWA is, and should be, non-prescriptive; this great strength of the approach, it flexibility, may also be perceived as one of its weaknesses; it is often perceived as being difficult to grasp. There is, therefore, perceived benefit in providing a more detailed description of one way of completing the process. Based upon the literature studied and the case studies completed, this book presents a detailed process for conducting a CWA; providing guidance to those new to the topic and hopefully consolidating the ideas of more experienced analysts. The presented approach is not intended to be the final word on the matter; it is readily acknowledged that the use of the framework and the choice of representation are likely to be heavily influenced by the domain and the required output of the analysis.

This book reinforces the importance of considering the framework in its entirety to gain an appreciation of the different constraint sets in existence in any given domain. It is contended that analyses that do not consider a range of constraint sets, captured within the framework, present a less complete model of the environment. Whilst discussing the problems of automation, Norman (1990) makes the following statement, which appears to be particularly applicable at this juncture:

> ...in design, it is essential to examine the entire system: the equipment, the crew, the social structure, learning and training, cooperative activity, and the overall goals of the task. Analyses and remedies that look at isolated segments are apt to lead to local, isolated improvements, but they may also create new problems and difficulties at the system level. (Norman, 1990 p.2)

In summary, this book presents a detailed description of how CWA can be applied to the analysis, design and evaluation of complex sociotechnical domains. The approach taken reinforces the importance of considering the framework in its entirety and provides a more complete description of how links between the relevant phases can be exploited.

Why CWA for Command and Control?

As the subsequent chapters in this book will discuss, the complexity embodied in command and control systems present significant challenges for modelling and analysis. Traditional reductionist modelling techniques, which decompose activity into a set of task sequences, can rarely be extended beyond stable and repeatable systems. An analysis or modelling technique is, therefore, required that can handle this inherent complexity and adaptability. One means of achieving this is to concentrate on the constraints shaping the way work is conducted in a given domain. In order to achieve this, it is advantageous to start by considering the environment independent of technology. As Moore's (1965) law tells us, computing power doubles every 18 months bringing with it new capabilities and products. If we do not consider our models in a technologically agnostic way, it will soon become significantly outdated and, in turn, unrepresentative.

The question is not: how do we use new technologies to execute our current tactics and doctrine better? It is instead: how might the new technologies enable us to do things differently? Without new thinking, the new technologies are likely to increase efficiency in a way that is largely arithmetic and incremental in nature. They enable us to execute today's tactics and operations in a quantifiably better way, but they still leave us with traditional, tightly controlled, and synchronized operations that are hierarchically planned and executed. (Smith, 2003 pp 74–75)

The domains within command and control are, clearly, both complicated and dynamic: the constituent parts of the system (including humans) are required to frequently change and adapt in time with the environment. Naikar (in press) points out that by focussing on constraints, rather than on particular ways of working, it is possible to support workers in adapting their behaviour online and in real time in a variety of situations, including unanticipated events. For these complex domains, an approach is required that models the conditions framing formative behaviour, allowing the examination of emergent, unpredicted, unanticipated actions. Cognitive Work Analysis (CWA) is presented as an approach that is particularly amenable for this domain. As previously stated, this book aims to address how CWA can be used to support the analysis, development and evaluation of complex sociotechnical systems. Due to their inherent complexity, specific attention will be placed on Command and Control systems. Through its focus on the system constraints, and its initial drive to be technologically agnostic, CWA supports revolutionary as well as evolutionary design. It is contended that; with formative thinking, based on approaches such as CWA, the exponential improvements described by Smith (2003) can be realised.

At this point in the book, it is important to temper the enthusiasm for the approach by acknowledging its limitations. The approach requires specific training for its application and understanding; the level of which will be explored later in the book (see the appendix to this document). The technique relies on the skill of the analyst to appropriately set boundaries for the analysts as well as make constant judgment calls to determine the level of the analysis. For these reasons, the outputs of the technique can vary significantly dependant on the skill of the analyst. As Chapter 3 will discuss, design is, essentially, an art dependant on creativity and the application of design principles. However, Chapter 3 will also show that the number of constraints, and the inherent flexibility, in these complex sociotechnical systems requires a structured approach. No approach exists, and probably never will exist, that can create an explicit link between problem and solution for complex sociotechnical systems. What this book sets out to demonstrate is that, through a series of developments, CWA can provide a structured framework for the modelling of constraints. This model of constraints forms the basis for the support of analysis, development and evaluation, provide designers and analysts with a common language and structure for the discussion and development of such systems.

Structure of the Book

This book has been constructed so that readers new to this specific subject area can read it linearly. An attempt has also been made to construct the individual chapters and case studies so that they can be read non-linearly, or independently of the early chapters, by those well versed in CWA with an interest in the extensions and proposed developments documented in this book. This means that there is some repetition in the description of CWA, in an attempt to create a resource for experts.

2   It's a Complex World

The book commences with a scene-setting chapter explaining in more detail the challenges encountered by both analysts and designers when they are faced with complex sociotechnical systems, further justifying the need for this research. The CWA framework is introduced and the need for formative behaviour is explored in detail.

3   Interaction Design

This chapter introduces the basic principles of interaction design to the reader; placing particular attention on interface design. The position of design on the art-science continuum is explored, as is the requirement for a structured approach to support design for complex sociotechnical systems. The intension of this chapter is not to provide an exhaustive guide to interface design; rather, the aim is provide the reader with context for latter phases where design is applied. A case study of a battlefield-planning tool is used to highlight the pertinence of some of the discussed design considerations.

4   Application of CWA in Familiar Domains

In order to illustrate the CWA framework and principles, introduced in Chapter 2, a 'simple' familiar artefact is used. The examples, in this and subsequent chapters, have been selected because they convey a particular message, the individual examples are not intended to stand alone as perfect examples of how to complete a CWA, most are incomplete in either analysis, design or evaluation; however, they each convey a different message; each explaining a different part of the technique. It is in keeping with the non-prescriptive, formative nature of the approach that a conscious decision has been made to present a wider breadth of examples rather than fewer overly detailed cases.

5 Applications of CWA in a Complex World

In this chapter the CWA framework is applied to complex command and control case studies. Two examples are selected that in turn address, analysis, the extraction of design recommendations and system evaluation. The first example presented analyses a military rotary wing mission planning software tool. The second investigates battlefield-management and addresses the evaluation of the impact of new technology introduced to a domain. The chapter explores, in detail, the benefits of applying the entire CWA framework; further, a new, theoretically grounded, approach to system evaluation is presented.

6 Using CWA to Design for Dynamic Allocation of Function

This chapter explores the social organisational and cooperation phase of the framework in the context of a dynamic command and control paradigm. The chapter starts with a thorough analysis of the domain; this analysis is then used to inform the design and development of a series of interfaces; the rapidly reconfigurable interfaces are designed to support dynamic allocation of function. The developed interfaces are designed to be role-specific modelling the system at a functional and physical level.

7 Designing Interfaces Using CWA

Much of the criticism for CWA and other human factors techniques are formed around the difficulty analysts have in taking analysis products and applying them to create design solutions or design recommendations. This chapter is structured with the specific aim of redressing this concern; this chapter explores the benefits of considering complex military planning within the Abstraction Hierarchy framework. The example described discusses the development of a military decision-support software tool. Here, the evaluation approach introduced in Chapter 5 is extended to consider the resultant high order effects of manipulations to objects at low levels of abstraction. Key question relating to interface design are investigated though systematic experimentation.

8 Development of a CWA Software Tool

In order to expedite the process of analysis, development and evaluation, as well as the training of novices: The development of a CWA software tool is discussed. The chapter addresses the need for the tool as well as cataloguing its development.

The benefits of the tool are compared against the paper-based approach along with standard graphics software.

9   *Does the Tool Make the CWA Process any Quicker or Easier?*

A military example is applied to evaluate the tool proposed in Chapter 8. The example is taken from a training video of a Battlegroup in the quick attack formation. From the analysis, additional requirements are extracted and discussed.

10 *Conclusions*

The final chapter concludes the study with discussion of the aims and objectives. Areas for further research are identified along with the key contribution to knowledge brought about by the research discussed in this book.

Appendix   Can it be Taught?

A key factor in the success of any developed approach is its ability to be learnt and assimilated by others. The appendix asks the questions 'can CWA be taught?' and 'how effective can the tuition be within a limited time period?' An experiment was conducted using final year design students to assess how learnable Work Domain Analysis is. The findings of this experiment informed the development of a CWA desktop software tool.

Chapter 2
It's a Complex World

Introduction

The aim of this chapter is to set the scene for the book, first the concept of complexity is introduced; this widely used term is explained in relation to sociotechnical systems. The specific example of command and control is used to highlight this. Cognitive Work Analysis is then presented as a framework to model these complex sociotechnical systems.

Complexity

Dependant on the context, the term complex has different meanings, whilst in mathematics it is used to describe a particular set of numbers, in this case we are considering complex by its definition of 'something with lots of interrelated parts'; a system. According to Walker et al (2007) at a fundamental level, complexity relates to the amount of information needed to describe a phenomenon under analysis. The closer that the phenomenon under analysis approaches complete randomness, the more data is needed until it 'cannot be described in shorter terms than by representing the [phenomenon] itself' (Bar Yam, 1997). However, 'something is complex if it contains a great deal of information that has a high utility, while something that contains a lot of useless or meaningless information is simply complicated' (Grand, 2000, p. 140 cited in Bar Yam, 1997). Complexity is not a binary state of either 'complex' or 'not complex', the level of complexity lies on a non-numerical scale, in the past people have attempted to quantify this through a set of heuristics, Woods (1988) proposes the following four dimensions as a framework for assessing the complexity of a system:

- Dynamism of the system: To what extent can the system change states without intervention from the user? To what extent can the nature of the problem change over time? To what extent can multiple on-going tasks have different time spans?
- Parts, variables and their interconnections: The number of parts and the extensiveness of interconnections between the parts or variables. To what extent can a given problem be due to multiple potential causes and to what extent can it have multiple potential consequences?
- Uncertainty: To what extent can the data about the system be erroneous, incomplete, or ambiguous – how predictable are future states?

•   Risk: What is at stake? How serious are consequences of users' decisions?

Highly complex systems can be characterised by all, or most, of the four dimensions identified by Woods (1988). These domains are classified by Rittel and Webber (1973) as 'Wicked problems', wicked problems are not fully understood, they have fuzzy boundaries, lots of stakeholders and lots of constraints with no clear solution. From a user-centred perspective, these 'wicked problems' present a significant challenge for designers. When a user is left to cope with complexity systems are harder to learn, understand and to use. They are likely to be unintuitive even for experienced users, often resulting in: high workload; reduction in tempo, efficiency and flexibility; and an increase in potential errors.

In the development of complex sociotechnical systems, it is the goal of the designer to design and develop a system that is usable, can be learnt in a reasonable amount of time, and can be understood so as to easily diagnose system errors and abnormalities. Tesler's Law of conservation of complexity tells us that some complexity is inherent in every process. There is a point beyond which you cannot simplify a process any further; you can only move the inherent complexity from one place to another. Essentially, what this means is that we cannot remove complexity, but we can transfer it to an automated action. An example of this transfer of complexity can be seen in automatic route finders on either online websites or in-car satellite navigation systems. The complex task of deciding the optimum route is not removed, instead the computer carries out the calculations required rather than the human user. Hollnagel (1995; 1992) reinforces this point, commenting that complexity cannot be removed, only hidden and to hide complexity is risky. An alternative approach is to design for complexity and provide the users with aids for coping with the complexity.

As stated earlier, one of the dimensions defined by Woods (1988) for describing complex systems is the number of constituent parts. This book will address the interaction of these constituent parts, paying particular attention to, how humans collaborate and perform within these sociotechnical systems. The term sociotechnical can be broken down into two composing parts: socio relating to the social subsystems and technical relating to technical subsystems. With this simple definition, it is clear that there are many systems in this category. According to Walker et al (2007), the phrase 'sociotechnical' is ubiquitous in human factors literature. As a 'theory', it describes the interrelatedness of socio and technical elements of a work domain and the non-linearities that are created when one or other is optimised. The conceptual response represented by Sociotechnical Theory is to shift the paradigm away from complex bureaucratic organisations 'doing' simple tasks, to instead, consider simpler organisations with more complex and, for the people carrying them out, more meaningful tasks. Sociotechnical theory holds the notion that the linkage between socio and technical parts is just as important as the 'parts' themselves. In other words, instead of outputs representing the sum of their socio and technical inputs, any such effects become multiplicative and emergent.

Command and Control C2

In basic terms Builder, Bankes and Nordin (1999) define command and control as:

> The exercise of authority and direction by a properly designated [individual] over assigned [resources] in the accomplishment of a [common goal]. Command and control functions are performed through an arrangement of personnel, equipment, communications, facilities, and procedures which are employed by a [designated individual] in planning, directing, coordinating, and controlling [resources] in the accomplishment of the [common goal]. (Builder, Bankes and Nordin, 1999, p.11)

The product of combining command (authority) with control (the means to assert this authority) are the emergent properties of, 'unity of effort in the accomplishment of a [common goal]' (Jones, 1993) and, 'decision superiority' (DoD, 1999, p.28). Despite the militaristic undertones, the notion of command and control is generic and not specific to a particular domain. Beyond the descriptive level, command and control is, by definition, a collection of functional parts that together form a functioning whole. Command and control is a mixture of people and technology (a sociotechnical system), typically dispersed geographically. Command and control domains are indeed sociotechnical systems of enormous complexity. As Storr (2005) points out 'Armed forces are astonishingly complex' according to Zsambok and Klein (1997) the battlefield can be characterised as an environment that has 'high stakes; is dynamic, ambiguous, and time stressed; and in which goals are ill defined or competing'. When compared against the dimensions proposed by Woods (1988) it is clear that we are dealing with a 'wicked problem'. Brehmer (2007) points out that:

> It is important to realize that, from a design perspective, the form of the C2 system is a normative concept; it specifies how C2 should be performed according to the design. There is, of course, no guarantee that C2 will be performed in that way. It is therefore important to distinguish between the form of the system, which specifies how C2 should be performed, and the process of C2 and the 'command culture', which denotes how it actually is performed. (Brehmer, 2007, p.214)

It is unsurprising that around the globe different defence organisations are taking significant interest in these domains, investigating how the introduction of new emergent technology can support them. As new technologies are introduced into these domains the number of components and the possibilities for interaction increases; according to the heuristics identified by Woods (1988), this makes the systems become ever more complex. The management of this complexity is of great importance; for the first time high-level commanders have the opportunity to access real time information about almost all of their assets (some times referred to as the long handle screwdriver effect). According to Lintern (2006), information overload has become a critical challenge within military operations; the problem is not so much one of too much information but of abundant information that is poorly organised and poorly represented. The user has the challenge of searching the information space to find, distinguish, summarise, integrate and understand the meaningful elements that can make a difference. Lintern (2006) goes on to say that information management has emerged as a significant contemporary challenge in modern warfare. The advantage now goes not to those with the more potent weaponry but to those with the more effective information system.

Research in this area generally falls under the title of 'Network Enabled Capability' (NEC) in the UK; this is, more or less, synonymous with what is term 'Network-Centric Warfare' (NCW) in the USA. Regardless of the title each of these initiatives is concerned with embracing new technologies to not only do things better, but, to do better things. By facilitating network capabilities, the distribution of information is made easier and, more importantly, significantly quicker. With this new technology, many of the previous physical constraints within a command and control domain have been removed; organisational structures can be rethought, as hierarchical structures are no longer the only efficient way to disseminate orders and information. According to Bruseberg and Lintern (2007), the advantages of NEC include aspects such as; the speed of information transfer over large distances and between many users; better access to information previously unavailable (e.g. by drawing on new sensor technologies); remote control of weapon systems; automated information processing aids; fast coordination of activities between diverse partners; and high-power data-storage aids. The challenge is in the implementation of these systems and the representation and structuring of this new information. An approach is clearly required that can handle this complexity and assist in this process.

What Makes CWA Different From Other Human Factors Methods?

When people think of human factors and design they generally think of User Centred Design (UCD), the adoption of UCD principles guides engineers to explicitly consider how users interacted with the products and services they are creating. As the term suggests 'User Centred Design', specifically participatory design, places a strong emphasis on the user, this means that, in most cases, the feedback from the user (time, error, performance data as well as subjective opinions) steers the direction for the design and development of the product. In many consumer products, this is a sensible approach; however, in highly complex sociotechnical systems this approach has limitations. When the system is considered in its entirety, individual users often have an incomplete understanding; their particular understanding of the system is likely to be biased to their particular role. Individual users often develop misconceptions of the constraints governing the system outside of their control. User opinion might not be impartial; it is not uncommon for users to be resistant to new revolutionary ideas.

As previously discussed, rather than describing existing behaviour, CWA offers a formative approach by focusing on possible behaviour. CWA also starts by focusing on system description that is independent of any actor; conversely, normative approaches, by focusing on existing activity, are more likely to be more individualistic. Normative approaches are also likely to be far more prescriptive focusing on the final product or solution, whereas, CWA aims to describe the initial conditions for behaviour to support the system in adaptation. This makes normative approaches more applicable for linearity and bureaucratic systems, whereas, CWA is much better equipped to analyse non-linear systems supporting emergent behaviour.

CWA was originally developed at the Risø National Laboratory in Denmark (Rasmussen, 1986) for use within the nuclear power industry. The technique was developed as a result of the electronic departments identified need for 'design for adaptation', this need to design for new situations was determined from a study of industrial accidents; the Risø researchers found that most accidents began with non-routine operations. According to Fidel and Peijtersen (2005), CWA's theoretical roots are in General Systems Thinking, Adaptive Control Systems, and Gibson's Ecological Psychology. CWA is particularly appealing as it can be applied in both closed systems, in which operations are predictable and options for completing a task are normally limited, and open systems, in which task performance is subject to influences and disturbances that cannot always be foreseen. The approach is also describes as being amenable for the design and development of systems that are not currently in existence (first-of-a-kind systems).

The CWA framework has been developed and applied for a number of purposes, including: system modelling (e.g. Hajdukiewicz, 1998); system design (e.g. Bisantz et al, 2003); training needs analysis (e.g. Naikar and Sanderson, 1999), training program evaluation and design (e.g. Naikar and Sanderson, 1999); interface design and evaluation (Vicente, 1999); information requirements specification (e.g. Ahlstrom, 2005); tender evaluation (Lintern and Naikar, 2000; Naikar and Sanderson, 2001); team design (Naikar et al, 2003); the development of human performance measures (e.g. Crone et al, 2003, 2007; Yu at al, 2002); and error management strategy design (Naikar and Saunders, 2003).

These applications have taken place in a variety of complex safety critical domains including: air traffic control (e.g. Ahlstrom, 2005); automotive (e.g. Jenkins et al, 2007d); aviation (e.g. Naikar and Sanderson, 2001; Xu, 2007); health care (e.g. Burns et al, 2006; Watson and Sanderson, 2007); hydropower (e.g. Memisevic et al, 2005); nuclear power (e.g. Olsson and Lee, 1994); naval (e.g. Bisantz et al, 2003); manufacturing (e.g. Higgins, 1998); military command and control (e.g. Jenkins et al, 2008a; Salmon et al, 2004); petrochemical (e.g. Jamieson et al); process control (e.g. Vicente, 1999); rail (e.g. Jansson et al, 2006) and road transport (e.g. Salmon et al, In Press)

According to Wong et al (1998) CWA was originally used to analysis processes control-type, physically coupled, causal system domains to identify structural associations between the physical components of a system and the abstract functions of the system. Wong et al (1998) go onto say that the approach appears not suited for use in human-activity-based intentional systems domains. Hajdukiewicz et al (1999) strongly contest this claim; they use an example of an ambulance dispatch system and a military command and control system to illustrate this. Hajdukiewicz et al (1999) contend that the model is well suited for intentional systems, they go onto suggest that Wong et al (1998) may have come to their conclusion as a result of an inappropriate scope and system boundary.

The first stage of CWA, work domain analysis, starts by considering how the system might reasonably perform (formative modelling), rather than focusing on how the system should perform (normative modelling), or how the system currently performs (descriptive modelling). This formative approach leads to an event and time independent description of the system (Sanderson, 2003; Vicente, 1999). The difference between normative and formative systems can be further highlighted with the example of a map. When driving to visit an unfamiliar place many people use web-based digital route planners to get directions. There are a number of websites available that allow the user to type in the post-code (or zip code) of the start and end-point to generate directions. Dependant on the website, the selected output may be normative, descriptive or formative. The normative solution would be a textual description (take the 2nd left, then the 3rd right). This can be the simplest way of getting there; however, if you have ever been lost with these kinds of directions, you will soon realise that the description becomes useless once context is lost and that a map (the formative model) is what is needed to get to the destination. Vicente highlights the problems of methods that do not allow the worker to work at a formative level, stating that:

> Workers do not, cannot, and should not consistently follow the detailed prescriptions of normative approaches. (Vicente, 1999, p.94)

The fundamental difference between normative task analysis techniques such as Hierarchical Task Analysis (HTA; Annett, 1971, 1996, 2004) and CWA is that HTA is focused on the hierarchical subdivision of goals, whereas, CWA focuses on the purpose and constraints of the domain that tasks and goals take part within. According to Miller and Vicente (2001), HTA's are formed by 'action' means-ends links. The method focuses on what actions need to be carried out to achieve a higher-level goal. CWA represents 'structural' means-ends relationships, illustrating the structural degrees of freedom of the system available to achieve higher-level purposes. Naikar et al (2005) describe the differences between CWA and task analysis as significant. Task analysis is event dependent because trajectories of behaviour can only be defined for specific situations with known goals and work requirements. Whilst it is possible to generate alternative 'plans' in a HTA, these plans can only be generated for known events. Naikar et al (2005) go on to point out that task analysis can only specify the information or knowledge that workers need for dealing with routine or anticipated situations. In contrast, by focussing on the analysis of constraints, CWA offers an event independent approach to work analysis that can identify the information or knowledge that workers need for dealing with a wide variety of situations, including novel or unanticipated events. Annett (2004) contradicts this view, in part, stating that HTA seeks to represent system goals and plans rather than focusing solely on observable aspects of performance; it is often difficult to contrast these different approaches as both have been significantly extended.

CWA is a formative approach; whereas, HTA seeks to fit somewhere between descriptive and normative. Sub-goals at the bottom of the hierarchy in an appropriately constructed HTA will mainly be descriptive, moving up the hierarchy they will become normative. Bruseberg (2005) addresses further limitation of HTA stating that HTA struggles with suitable representations for parallel and iterative processes (e.g. managing of conflicting goals; review of own approach; maintaining social interaction; retrieval from memory, prioritisation, storage, and forgetting). Such processes cannot be captured accurately by a hierarchical structure, since they involve overarching and overlaying activities that apply across the functional goal structures described by HTA. CWA performs this task far more effectively as it focuses on the functions and constraints of the system. CWA often requires a more disciplined approach, according to Naikar et al (2006), focusing on tasks is more natural to analysts and subject matter experts, so in order to remain thinking in terms of functions they are forced to examine the reason for which the tasks were being performed. Naikar and Lintern (2002) comment on the problems of the normative work analysis approach in complex sociotechnical situations, stating that; when temporally ordered actions, that should be done, are specified it results in workers being ill prepared to cope with unanticipated events. Descriptive work analysis approaches often converge on patterns of behaviour that are more robust and economical than normative analysis; however, these approaches can only deal with situations that have been anticipated. It is natural human behaviour to develop short cuts and work-around that move away from the normative approach. In many cases, the normative approach is used in industrial action as a way of making systems more cumbersome and reducing productivity (the work-to-rule action). According to Naikar (2006b), CWA recognises that many tasks in complex sociotechnical systems are discretionary and that workers have a great variety of options with respect to, what to do, when and how. The importance of a system ability to handle unexpected events cannot be expressed enough. As Don Norman (2007) points out:

> We know two things about unexpected events: first; they always occur, and second, when they do occur, they are always unexpected. (Norman 2007, p.13)

Naikar and Lintern (2002) argue that there are three thematic concerns discussed in Vicente's (1999) description of CWA. These include that complex sociotechnical systems cannot be anticipated completely; because of this, workers must be adaptive. The second theme is that the technology itself offers radical new approaches to work. Naikar and Lintern (2002) go on to comment that designers fail to take advantage of many new opportunities because they are caught in an evolutionary task-artefact cycle (Carroll, 1991; see Chapter 1) in which existing work practices are allowed to constrain the options for new designs. The third theme proposed by Naikar and Lintern (2002) is that in many other forms of work analysis, there is a tendency to focus exclusively on the human cognitive system, whereas, CWA attempts to also address the physical and social constraints. CWA is separated from many of the other methods, or frameworks, available to Human Factors practitioners because:

> rather than prescribing how work should be done or describing how it is currently being done, CWA seeks to identify how work could be done if the appropriate tools were made available. (Vicente, 1999, p.340)

Because of the industrial and digital revolutions, there was a zealous drive to automate much of the activity performed by humans. As a result, many process-driven activities have been completely reduced to a set of algorithms or rules and automated. As Bainbridge (1987) was keen to point out, automation is often misconceived as a means of improving a system. Although automation is often identified as a major culprit in industrial accidents, Norman (1990) proposes that the problems result from inappropriate application, not the commonly blamed culprit of 'over-automation'. According to this view, operations would be improved either with a more appropriate form of automation or by removing some existing automation.

Many activities in complex sociotechnical domains require workers to engage in open-ended and creative, discretionary decision-making (Vicente, 1999, p26). Norman (1990) points out, that what is needed is a soft, compliant technology, not a rigid, formal one. According to Bruseberg (2005), the main underlying premise of CWA is that one cannot understand task approaches, and therefore cognition, without understanding the nature of the work domain. Complex sociotechnical systems are rarely consistent with time; they are forced to change as new unexpected and unanticipated events take place. Naikar et al (2005) comment that, as the system is dynamic, the goals and work demands are frequently changing. CWA is fundamentally concerned with designing systems for adaptation; this is achieved by focusing on the overall purpose and the constraints of the system, both physical and social.

From examination of Table 2.1 it appears as if CWA fits into the definition of a system design. The philosophy behind user centred design is that the user knows best. The philosophy is best suited to when users understand their goals, needs and preferences, in many simple, generally single artefact based-systems, that sit outside the 'complex sociotechnical system' description this design approach may be well suited; however, for complex systems there are few users that can interpret and articulate the required information to engage in the design activity. Activity-Centred Design as the name suggests is focused on the activity taking place rather than the actor performing it, in many cases this approach looks at removing tasks

Table 2.1   Four approaches to design (reproduced from Saffer, 2007)

Approach	Overview	Users	Designer
User-Centred Design	Focuses on user needs and goals	Guide the design	Translate user needs and goals
Activity-Centred Design	Focuses on the tasks and activities that need to be accomplished	Perform the activities	Creates tools for actions
Systems Design	Focuses on the components of a system	Set the goals of the system	Makes sure all the parts of system are in place
Genius Design	Relies on the skill and wisdom of designers used to make products	Source of validation	Is the source of inspiration

from users to automate them, the limitation of this technique and of user centred design is that they both fixate on a particular task. Saffer (2007) offers an old adage that seems extremely pertinent:

> You get a different result if you tell people to design a vase than if you tell them to design something to hold flowers. By focusing on small tasks, designers can find themselves designing vase after vase and never a hanging garden. (Saffer, 200, p.357)

In systems design, individual users and activities are not discarded they are deemphasised; instead a much greater emphasis is placed on the system as a whole and the context in which it sits. As with most theories and frameworks semantics are crucial, it is important within CWA to understand the difference between goals and purposes. In simple terms, we conduct a task to achieve a goal, and in turn, we strive to achieve goals in order to enact the purpose of a system. Many authors of CWA literature are very keen to emphasise the difference between goals and purposes (Naikar et al, 2005, Burns and Vicente, 2001, Burns and Hajdukiewicz, 2004). Vicente (1999) offers definitions for both of these terms in the glossary of his book:

> Goal – A state to be achieved, or maintained, by an Actor at a particular time. Note that goals are attributes of Actors not Work Domains, and that they are dynamic. (Vicente, 1999, p.7)

> Purpose – The overarching intentions that a Work Domain was designed to achieve. Note that Purposes are properties of Work Domains, not Actors and that they are relatively permanent. (Vicente, 1999, p.9)

The implications of this distinction are explained by Naikar et al (2005):

> The purposes of a work system do not change from situation to situation nor do the physical resources that are available in the work system. Therefore, when confronted with novel or unanticipated events, workers can rely on their knowledge of the work-domain constraints to explore a variety of ways for dealing with the situation while remaining within the boundaries of acceptable performance. (Naikar et al, 2005, p6)

The Framework

According to Naikar (2006c), as there are several types of constraints that can shape workers' behaviour, several dimensions of analysis are necessary. In his early work, Rasmussen (1990) discusses the many applications of CWA as:

- The integration and development of cross disciplinary research theories in complex work domains.
- A methodology for design and specification of the functionality of integrated, complex, multimedia information systems.
- A methodological tool for planning field studies and data collection in various, actual work domains.
- A framework for comparison of the features of different work places with respect to behaviour shaping features.
- A method for evaluation of information systems and the behaviour of individual users and organisations for whom the systems are designed.

Vicente (1999) offers a more definitive description of the framework including abstraction hierarchies, decision-ladders, information flow maps and the Skills Rules Knowledge (SRK) framework. These tools can be seen in Figure 2.1. The technique is divided into five phases each focusing on different constraint sets, the commonly accepted representations for these phases are described on the right-hand side in Figure 2.1, also included are some guidance of possible acquisition methods for these phases.

It should be noted that depending on the literature studied, the phases of CWA may differ in name and slightly in their definition. Table 2.2 shows the difference between the phases proposed by Rasmussen (1994) and Vicente (1999). For consistency throughout the reminder of this text the phases shall be referred to in the convention suggested by Vicente (1999).

There is no prescriptive guidance for CWA; many analyses will not focus on all of the five phases, the majority of analyses tend to focus heavily on the first phase; Work Domain Analysis. The type of analysis conducted is likely to depict which of the phases are used, and in what ratio. Interface design, as described in 'Ecological Interface Design' (EID) (Burns and Hajdukiewicz, 2004), tends to use only the first stage Work Domain Analysis (WDA). An analysis for team design is more likely to

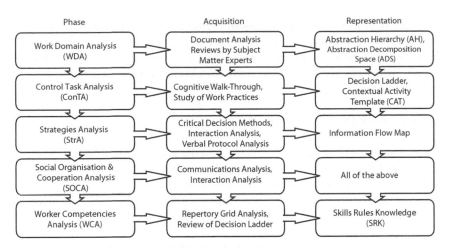

Phase	Acquisition	Representation

Figure 2.1 **The five phases of CWA according to Vicente (1999) (acquisition methods added from Lintern et al (2004))**

Table 2.2 **Comparison of phase definitions (reproduced from Naikar and Pearce 2003)**

Type of Constraints	Rasmussen at al (1994)	Vicente (1999)
Purposes priorities and values, general functions, and physical resources	Work domain analysis	Work domain analysis
Operating modes or work situations and work functions	Activity analysis in work domain terms	Control task analysis
Decision making functions or task control	Activity analysis in work decision making terms	
Strategies for making decisions or achieving control tasks	Activity analysis in terms of mental strategies	Strategies analysis
Distribution of work including allocation of work to individuals; organisation of individuals into teams; and communication requirements	Analysis of the work organisation	Social organisation and cooperation analysis
Generic human capabilities and limitations and competencies of workers (e.g. skills, attitudes)	Analysis of system users	Worker competencies analysis

focus on Control Task Analysis and Social Organisational and Cooperation Analysis. This flexibility of the suite of tools allows the CWA approach to be used throughout the system life-cycle for the design, development, representation and evaluation of both conceptual and operational systems. A brief overview of the five phases is given before presenting a more detailed description of CWA.

Phase 1 – Work Domain Analysis (WDA)

According to Naikar (2006b), WDA identifies the constraints on workers' behaviour that are imposed by the purposive and physical context, or problem space, in which workers operate. WDA is used to define the task environment; the environment that the activity is conducted in. WDA identifies a fundamental set of constraints on the actions of any actor, thus providing a solid foundation for subsequent phases.

The system domain is represented at a number of conceptual levels. At the highest level the system's *raison d'être* is represented; at the lowest level the physical objects within the system are described. The process of WDA addresses not only what is performed, but also, how and why.

Phase 2 – Control Task Analysis (ConTA)

Control Task Analysis addresses the constraints on activity imposed by specific situations. ConTA is used to understand the task; it allows us to identify the requirements associated with known, recurring classes of situations. This phase specifies the input and end goal without explicitly describing the process in the middle. The phase identifies what needs to be done independently of how or by whom.

Phase 3 – Strategies Analysis (StrA)

Strategies Analysis addresses the constraints influencing the way in which activity can be conducted. StrA looks at filling the 'black box' left in Control Task Analysis; it looks at different ways of carrying out the same task. Wherever the previous phase dealt with the question of what needs to be done this phase addresses how it can be done.

Phase 4 – Social Organisation and Cooperation Analysis (SOCA)

Social Organisation and Cooperation Analysis addresses the constraints imposed by organisational structures or specific actor roles and definitions. SOCA investigates the division of task between the resources and looks at how the team communicates and cooperates. The objective is to determine how the social and technical factors in a sociotechnical system can work together in a way that enhances the performance of the system as a whole.

Phase 5 – Worker Competencies Analysis (WCA)

Worker Competencies Analysis addresses the constraints dictating the possible actor behaviour within different situations. WCA investigates the behaviour required by the humans within the system required to complete tasks. Typically, this behaviour is modelled using Rasmussen's (1983) Skills, Rules and Knowledge (SRK) Taxonomy.

Framework for Conducting CWA

This section revisits each of the phases in turn, providing a more complete description. The phases are described from a theoretical basis. Examples of application can be found in Chapters 4 and 5, more detailed analyses are presented in Chapters 6 and 9.

Work Domain Analysis

WDA is the most commonly used component within CWA. The WDA is conducted at the functional, rather than behavioural level. Lintern (2005) describes WDA as a design artefact commenting that it organises information in a systematic manner that will support design. According to Burns and Hajdukiewicz (2004) the concept for WDA was developed at Risø National Laboratory as a result of observations of problem solvers, this lead to the realisation that people reasoned a problem using a structure basically asking how (how does this work) and why (why is this here). This was brought together to form means-ends links within a hierarchy, this hierarchy forms a basis for storing the relationships within a system.

The tool Vicente (1999) recommends for WDA is the Abstraction Decomposition Space (ADS). The ADS is a two dimensional space representing an Abstraction Hierarchy and a decomposition hierarchy. The ADS represents the entire domain under analysis outlined by an analyst-defined boundary. Along the decomposition axis the table shows the entire system under examination (total system), stepping down through levels of detail to a component level. The abstraction component of the diagram models the same system at a number of levels of abstraction; at the highest level the overall functional purpose of the system is considered, at the lowest level the individual components within the system are described. Generally, five levels of abstraction are used; these levels have been summarised in Table 2.3 by Naikar et al (2005). The old labels refer to the original labels generated by Rasmussen and used by Vicente (1999) these were subsequently updated. To avoid confusion for the remainder of this document the 'New Labels' have been adopted as a standard.

An Abstraction – Decomposition matrix is an activity independent description of a work domain. The ADS focuses on the purpose and constraints of the work domain, and for this reason analyses are typically performed with data collected

Table 2.3   Descriptions of levels of abstraction

'Old Labels'	'New Labels'	Description
Functional Purposes	Functional Purposes	The purposes of the work system and the external constraints on its operation
Abstract Functions	Values and Priority Measures	The criteria that the work system uses for measuring its progress towards the functional purposes
Generalised Functions	Purpose-Related Functions	The general functions of the work system that are necessary for achieving the functional purposes
Physical Functions	Object-Related Processes	The functional capabilities and limitations of physical objects in the work system that enable the purpose-related functions
Physical Form	Physical Objects	The physical objects in the work system that afford the object related processes

from discussions with engineering experts, and with the review of design and engineering documents. This provides an understanding of how and why the structures work together to enact the desired purposes. Sharp and Helmicki (1998) point out that if the data sources are inadequate, the analysis will be correspondingly inadequate, but even partial and incomplete knowledge can be used to provide a helpful understanding of the work domain

Means-ends relationships form the basis of decision-making and are captured in the process for creating an ADS. These relationships capture the affordances of the system; the process involves examining what needs to be done within a system and understanding the available means for completing these ends. In complex work domains there are often many-to-many relationships, these require the user to select from a number of physical arrangements to meet a specific purpose. In the same way, one specific physical arrangement may be the only method for two independent purposes requiring the user to make decisions. In routine or highly familiar situations, the means-ends relations that must be considered by workers are usually well established and stable. Decision making in these situations is relatively straightforward and reflects rule-based reasoning (Rasmussen et al., 1994). Naikar et al (2005) explain this well using an example of the home; people may know the effects of eating out more than once or twice a week on their savings so that making a decision about whether to eat out or not on any particular occasion may be relatively straightforward (rule-based). An unfamiliar decision that will require more thought; inhabitants may need to consider explicitly all possible means-ends relations in deciding whether to accommodate an elderly parent at home or in a nursing home. Decision-making in these situations can be relatively challenging or demanding and is characterised by knowledge-based reasoning (Rasmussen et al., 1994).

According to Naikar and Sanderson (1999), when a work domain is a predominantly physical one, such as process control, it us usually straightforward to identify structural means-ends functions and relations. However, when a work domain s more intentional in nature (its structural properties emerging from human intentions) it can become more difficult to cleanly separate structural and action means-ends relations. These socially, rather than physically, constructed relationships take considerably more analytical effort to preserve the distinction from action means-ends.

The Abstraction Hierarchy allows the system to be considered at a number of levels, existing systems are likely to be modelled from the bottom up, whereas, new systems are more likely to be modelled from the top down. The hierarchy can be used as a structure to think about existing systems. Using a top down approach, a system can be considered in a technologically agnostic way, allowing the analysts to consider what the systems is trying to do rather than what it actually is doing. Using a bottom up approach diversification can be investigated by considering new purposes for existing systems.

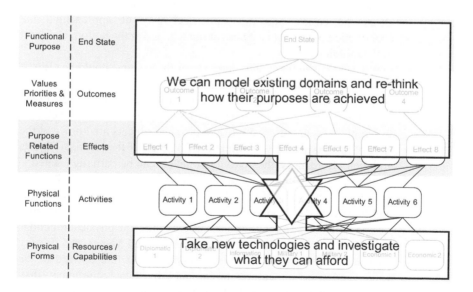

Figure 2.2   Abstraction Hierarchy showing top down and bottom up concepts

Naikar et al (2005) detail a nine step methodology for completing an ADS, this nine step list can be explained in detail by supplementing it with information from Burns and Hajdukiewicz's (2004) hints for completing a WDA. This nine steps approach is applied to a military scenario later in the book in Chapter 9.

1. Establish the purpose of the analysis. At this point it is important to consider what is sought, and expected, from the analysis. This will influence which of the five phases are used and in what ratio.
2. Identify the project constraints. Project constraints such as time and resources will all influence the fidelity of the analysis.
3. Identify the boundaries of the analysis. The boundary needs to be broad enough to capture the system in detail; however, the analysis needs to remain manageable. It is difficult to supply heuristics, as the size of the analysis will in turn define where the boundary is set.
4. Identify the nature of the constraints in the work domain.
5. Identify the sources of information for the analysis. Information sources are likely to include, among others: Engineering documents (if the system exists), the output of structured interviews with SMEs and the outputs of interviews with stakeholders. The lower levels of the hierarchy are most likely to be informed from engineering documents and manuals capturing the physical aspects of the system. The more abstract functional elements of the system are more likely to be elicited from stakeholders or literature discussing the aims and objectives of the system.
6. Construct the ADS with readily available sources of information.
   - Often the easiest and most practical way of approaching the construction of the AH is to start at the top, and at the bottom and meet in the middle.
   - In most cases, the functional purpose for the system should be clear, it is the reason that the system exists.
   - By considering ways of determining how to measure the success of the functional purpose(s) the values and priority measures can be set.
   - In many cases, at least some of the physical objects are known at the start of the analysis. By creating a list of each of the physical objects and their affordances, the bottom two levels can be partially created.
   - Often, the most challenging part of an AH is creating the link in the middle between the physical description of the systems components and functional description of what the system should do. The purpose related functions level involves considering how the physical functions can be used to have an affect on the identified values and priority measures. The purpose related functions should link upward to the values and priority measures to explain why their requirement; they should also link down to the physical functions, explaining how they can be achieved.

- The next stage is to complete the means-ends links; checking for unconnected nodes and validating the links. The why-what-how triad should be checked at this stage.

7. Construct the ADS by conducting special data collection exercises. After the first stage of constructing the AH there are likely to be a number of nodes that are poorly linked; indicating an improper understanding of the system. At this stage, it is often necessary to seek further information on the domain from literature or from SMEs

8. Review the ADS with domain experts. The validation of the AH/ADS is a very important stage. Although this has been listed as stage 8 it should be considered as an iterative process. If it is possible to access the SMEs it is advantageous to consult with them throughout the process. Often the most systematic process for validating the AH is to step through node by node checking the language. Each of the present links should be validated along with each of the correct rejections.

9. Validate the ADS. Often the best way to validate the ADS is to consider known recurring activity checking to see that the AH contains the required physical objects and that the values and priority measures captured cover the function aims of the modelled activity.

Naikar et al (2005) have generated some very useful prompts to aid the analyst in completing the ADS. These prompts assist in the classification of nodes in to the specific levels of abstraction and decomposition. The prompts can be seen in Table 2.4 and Table 2.5.

Table 2.4   Naikar et al (2005) list of prompts for abstraction

	Prompts	Keywords
Functional Purposes	**Purposes** • For what reasons does the work system exist? • What are the highest-level objectives or ultimate purposes of the work system? • What services does the work system provide to the environment? • What needs of the environment does the work system satisfy? • What role does the work system play in the environment? • What has the work system been designed to achieve? • What are the values of the people in the work system? **External Constraints:** • What kinds of constraints does the environment impose on the work system? • What values does the environment impose on the work system? • What laws and regulations does the environment impose on the work system? • What societal laws and conventions does the environment impose on the work system?	**Purposes** Reasons, goals, objectives, aims, intentions, mission, ambitions, plans, services, products, roles, targets, aspirations, desires, motives, values, beliefs, views, rationale, philosophy, policies, norms, conventions, attitudes, customs, ethics, morals, principles. **External constraints:** Laws, regulations, guidance, standards, directives, requirements, rules, limits, public opinion, policies, values, beliefs, views, rationale, philosophy, norms, conventions, attitudes, customs, ethics, morals, principles.
Values and Priority Measures	• What criteria can be used to judge whether the work system is achieving its purposes? • What criteria can be used to judge whether the work system is satisfying its external constraints? • What criteria can be used to compare the results or effects of the purpose-related functions on the functional purposes? What are the performance requirements of various functions in the work system? How is the performance of various functions in the work system measured or evaluated and compared?	Criteria, measures, benchmarks, tests, assessments, appraisals, calculations, evaluations, estimations, judgements, scales, yardsticks, budgets, schedules, outcomes, results, targets, figures, limits. Measures of: effectiveness, efficiency, reliability, risk, resources, time, quality, quantity, probability, economy, consistency, frequency, success.

Table 2.4   *Continued*

	Prompts	Keywords
	• What criteria can be used to assign priorities to the purpose-related functions? What are the priorities of the work system? How are priorities assigned to the various functions in the work system? • What criteria can be used to allocate resources (e.g. material, energy, information, people, money) to the purpose-related functions? What resources are allocated to the various functions of the work system? How are resources allocated to the various functions of the work system?	Values: laws, regulations, guidance, standards, directives, requirements, rules, limits, public opinion, policies, values, beliefs, views, rationale, philosophy, norms, conventions, attitudes, customs, ethics, morals, principles.
Purpose Related Functions	• What functions are required to achieve the purposes of the work system? • What functions are required to satisfy the external constraints on the work system? • What functions are performed in the work system? • What are the functions of individuals, teams, and departments in the work system? • What functions are performed with the physical resources in the work system? • What functions coordinate the use of the physical resources in the work system?	Functions, roles, responsibilities, purposes, tasks, jobs, duties, occupations, positions, activities, operations.
Object Related Processes	• What can the physical objects in the work system do or afford? • What processes are the physical objects in the work system used for? • What are the functional capabilities and limitations of physical objects in the work system? • What physical, mechanical, electrical, or chemical processes are afforded by the physical objects in the work system? • What functionality is required in the work system to enable the purpose-related functions?	Processes, functions, purposes, utility, role, uses, applications, functionality, characteristics, capabilities, limitations, capacity, physical processes, mechanical processes, electrical processes, chemical processes.

Table 2.4   *Concluded*

	Prompts	Keywords
Physical Objects	• What are the physical objects or physical resources in the work system – both man-made and natural? • What physical objects or physical resources are necessary to enable the processes and functions of the work system? • What is the inventory (e.g. names, number, types) of physical objects or physical resources in the work system? • What are the material characteristics (e.g. external form including shape, dimensions, colour; internal configuration; material composition) of physical objects or physical resources in the work system? • What is the topography or organisation (e.g. layout or location of physical objects in relation to each other) of physical objects or physical resources in the work system?	Man-made and natural objects: tools, equipment, devices, apparatus, machinery, items, instruments, accessories, appliances, implements, technology, supplies, kit, gear, buildings, facilities, premises, infrastructure, fixtures, fittings, assets, resources, staff, people, personnel, terrain, land, meteorological features. Inventory: names of physical objects, number, quantities, brands, models, types. Material characteristics: appearance, shape, dimensions, colour, attributes, configuration, arrangement, layout, structure, construction, make up, design. Topography: organisation, location, layout, spacing, placing, positions, orientations, ordering, arrangement.

Table 2.5   Naikar et al (2005) list of prompts for decomposition

Prompts	Keywords
Levels of decomposition: • What is viewed as the whole system in the work domain? • What is the coarsest level at which workers view the work system? • What is the whole system around which work is organised in the work domain? • What do workers view as the parts of the work system? • What is the most detailed level at which workers view the work system? • What are the different levels of detail at which workers view the work system? • What are the parts around which work is organised in the work system? **Part-whole relations:** • Are the entities at higher levels of decomposition composed of the entities at lower levels of decomposition? • Are the entities at lower levels of decomposition parts of the entities at higher levels of decomposition?	Names of wholes or parts of: organisations, physical structures, physical spaces, conceptual structures, groups, teams, functions, positions, arrangements, aggregations, formations, assemblies, segments, pieces, units, components, systems, subsystems, divisions, branches, sectors, departments

Control Task Analysis

Control Task Analysis is used to understand the task; it allows the requirements associated with known, recurring classes of situations to be identified. Naikar (2006c) comments that, by identifying the activity that is necessary to achieve the objectives of a system, with a given set of physical resources Control Task Analysis complements WDA. This phase specifies the input and end goal leaving a black box in the middle. The phase identifies what needs to be done independently of how or by whom. According to Sanderson (2003), control tasks emerge from work situations and they transform inputs (e.g. current state, targets, etc.) into outputs (decisions, control actions, etc.).

Vicente (1999) recommends the use of the decision-ladder (see Figure 2.3) for this phase of CWA. The decision-ladder was developed by Jens Rasmussen who observed that expert users were relying on rule-based behaviour to conduct tasks. Rasmussen (1974) states that the sequence of steps between the initiating cue and the final manipulation of the system can be identified as the steps a novice must necessarily take to carry out the sub task.

The ladder can be seen to contain two different types of node: the rectangular boxes represent data-processing activities and the circles represent states of knowledge that result from data processing. According to Vicente (1999), the decision-ladder represents a linear sequence of information processing steps, but is 'bent in-half'. Novice users are expected to follow the decision-ladder in a linear fashion, whereas, expert users are expected to link the two halves by short-cuts (see Figure 2.3). According to Naikar and Pearce (2003), the left side of the decision-ladder represents the observation of the current system state, whereas, the right side of the decision-ladder represents the planning and execution of tasks and procedures to achieve a target system state. Sometimes observing information and diagnosing the current system state immediately signals a procedure to execute. This means that rule based shortcuts can be shown in the centre of the ladder. On the other hand, effortful, knowledge-based

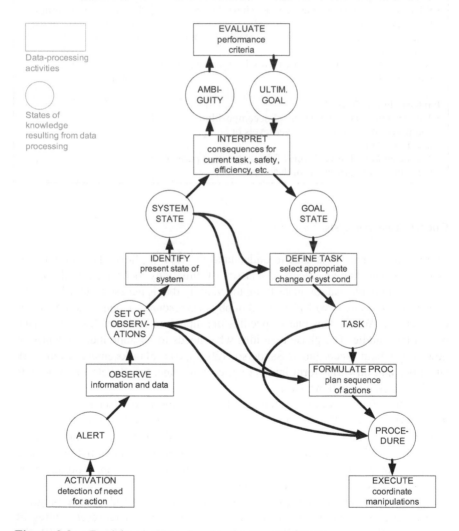

Figure 2.3   Decision-ladder showing leaps and shunts

goal evaluation may be required to determine the procedure to execute; this is represented in the top of the ladder. There are two types of shortcut that can be applied to the ladder; 'shunts' connect an information-processing activity to a state of knowledge (box to circle) and 'leaps' connect two states of knowledge (circle to circle); this is where one state of knowledge can be directly related to another without any further information processing. It is not possible to link straight from a box to a box as this misses out the resultant knowledge state. Cummings and Guerlain (in review) point out that when a shortcut is taken, various information-processing actions are bypassed but the desired results are still achieved. The decision-ladder not only displays these shortcut relationships in information-processing activities, it also highlights those states of knowledge that are bypassed if a shortcut is taken. According to Cummings and Guerlain (in review) the decision-ladder maps rather than models the structure of a decision-making process. In the case of systems with computer-based decision support tools, the decision-ladder represents the decision process and states of knowledge that must be addressed by the system whether or not a computer or a human makes the decision.

Based upon Rasmussen et al's (1994) definition of activity analysis in work domain terms Naikar et al (2006) have developed the Contextual Activity Template for use in this phase of the CWA (see Figure 2.4). This template is one way of representing activity in work systems that are characterised by both work situations and work functions. Work situations are situations that can be decomposed based on recurring schedules or specific locations. Rasmussen et al (1994) describes work functions as, activity characterised by its content

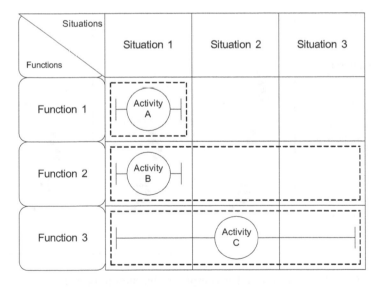

Figure 2.4   Contextual Activity Template

independent of its temporal or spatial characteristics (Rasmussen et al. 1994); these functions can often be informed from the middle level of the Abstraction Hierarchy, the purpose related functions. Rasmussen et al (1994) recommends that the analyst decompose on either work functions or work situations. Naikar et al (2006) plot these on two axes so that their relationship can be investigated allowing the representation of activity in work systems that are characterised by both work situations and work functions. The work situations are shown along the horizontal axis and the work functions are shown along the vertical axis of the Contextual Activity Template. The circles indicate the work functions with the bars showing the extent of the table in which the activity typically occurs. The dotted boxes around each circle indicate all of the work situations in which a work function can occur (as opposed to must occur); thus, capturing the constraints of the system. Figure 2.4 shows three activities plotted on the Contextual Activity Template. These activities can take place in three situations. Activity A can only take place in Situation 1; Activity B can take place in any of the situations although, typically, it only takes place in Situation 1; Activity C can take place in any situation and typically does.

As Figure 2.5 shows the decision-ladder can be plotted onto the Contextual Activity Template to add additional information for each possible combination of work situation and function (Naikar et al, 2006).

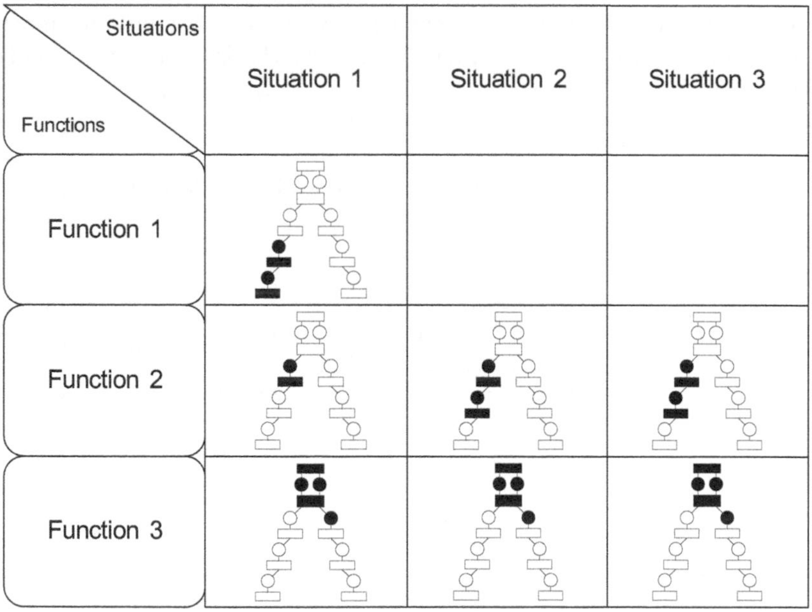

Figure 2.5   Contextual Activity Template (decision making)

Strategies Analysis

The strategy adopted under a particular situation may vary significantly. Different agents may perform tasks in different ways, and the same agent (either human or non-human) might perform the same task in a variety of different ways. Naikar (in press) comments that when an actor's work demands are low, actors may adopt a strategy that is cognitively more intensive, whereas, when their work demands are high, actors may adopt a strategy that is cognitively less intensive. The strategy that the actor chooses to select will be dependant on a huge number of variables, including, among others; their experience, training, workload, and familiarity with the current situation. To complicate things even further the same actor may select different strategies on different occasions.

Vicente recommends the use of information flow maps; according to Vicente (1999), information flow maps are used to describe the categories of cognitive task procedures that constitute strategies. These information flow maps can be very complex as in Figure 2.6, application is very case specific; with Vicente (1999) himself stating that the modelling tool has not reached the same level of maturity as the decision-ladder or the abstraction-decomposition space, because it has not been described as a generic tool. Naikar (2006c) supports this view commenting that Strategies Analysis has not yet been developed in a generic format but has only been created for specific applications.

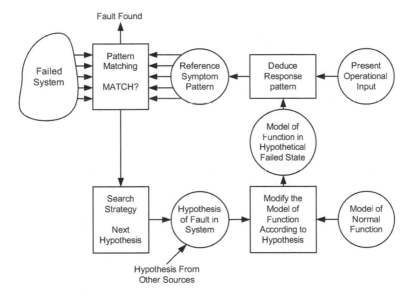

Figure 2.6   Information flow map for hypothesis and test search strategy (adapted from Vicente, 1999)

Ahlmstrom (2005) offers a simplified flow map, an example of this is shown in Figure 2.7; here the situation is broken down to a 'start state' and an 'end state', connecting the two, in the middle, are a number of strategies for the transformation.

Naikar (2006c) supports the adoption of structured flow maps commenting that Strategies Analysis is not concerned with defining detailed sequences of actions or mental processes. Instead, Strategies Analysis is concerned with identifying general 'categories of cognitive procedures'; which are idealised, abstract, descriptions of particular sequences of operations. Naikar (2006c) further discusses Strategies Analysis pointing out that; it recognises that workers will often switch between multiple strategies while performing a single activity in order to deal with task demands.

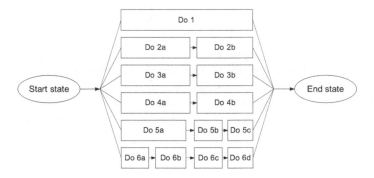

Figure 2.7   Strategies analysis

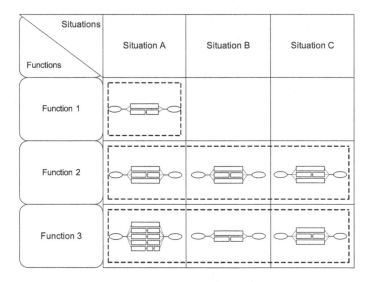

Figure 2.8   Strategies analysis plotted on the Contextual Activity Template

The Contextual Activity Template can be used as basis for eliciting the Strategies Analysis required. The matrix of functions and situations illustrates the possible activity within the domain (See Figure 2.8) to perform a complete analysis, a Strategies Analysis should be created for each of the cells where activities can take place (marked by the dotted line) the differences to the strategies available, effected by the situational constraints, can then be examined.

Social Organisation and Cooperation Analysis

Social Organisation and Cooperation Analysis (SOCA) addresses the division of tasks between the systems resources (both human and non-human). SOCA is concerned with how the team communicates and cooperates. The objective is to determine how the social and technical factors in a system can work together in a way that enhances the performance of the system as a whole. It is possible to map each of the actor types onto the existing tools in order to show who has the capability of doing what, thus providing a mechanism for informing decisions related to the allocation of function. Vicente (1999) does not specify a specific tool for use in this phase; instead it is suggested that an amalgamation of the tools already discussed should be used. By using the existing framework and mapping onto them the roles and responsibilities of sub-teams or individuals, teamwork can be better understood. Using shading it is possible to colour the existing products developed in the first three phases to show where each of the actor groups can conduct tasks.

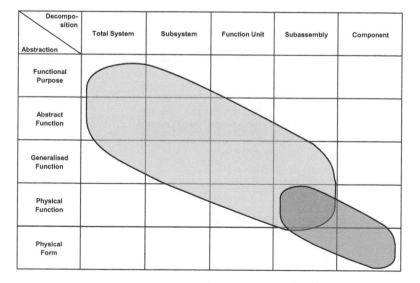

Figure 2.9   Example of two actors mapped on to the ADS

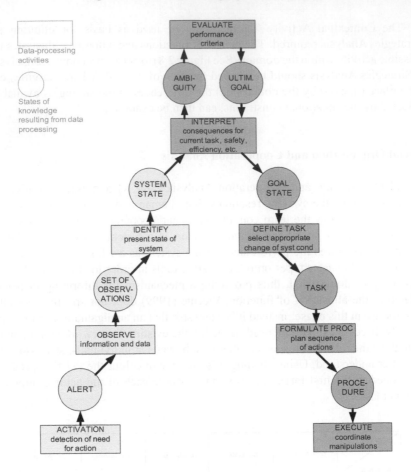

Figure 2.10   **Example of two actors mapped on to the decision-ladder**

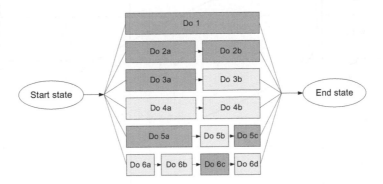

Figure 2.11   **Example of two actors mapped on to the Strategies Analysis flow diagram**

SOCA recognises that organisational structures in many systems are generated on line and in real time by multiple, cooperating actors responding to the local context (e.g. Beuscart, 2005). In the words of sociotechnical theory this would be a demonstration of the autonomy granted to groups and the freedom members of a group have to regulate their own internal states; relating themselves to the wider system. SOCA is therefore, expressive of the 'simple organisation/complex job' philosophy. It is not necessarily concerned with planning upfront the nature of organisational structures that should be adopted in different situations. It is instead concerned with identifying the set of possibilities for work allocation, distribution and social organisation. SOCA explicitly aims to support flexibility and adaptation in organisations (the sociotechnical principle of 'equifinality'; Bertalanffy, 1950) by developing designs that are tailored to the requirements of the various possibilities (the sociotechnical principle of 'multifunctionality'; Cherns, 1987). Ironically, SOCA is one of the more neglected phases of CWA (most emphasis being given to Work Domain Analysis).

Naikar (2006c) discusses the application of SOCA beyond the description of Vicente (1999) and Rasmussen (1994); applying it to the design of new organisational structures. Naikar (2006c) argues that SOCA recognises that in many systems, flexible organisational structures that can be adapted to local contingencies are essential for dealing with unanticipated events. Rather than defining a single or best organisational structure, SOCA is concerned with identifying the criteria that may shape or govern how work might be allocated across actors. Naikar (2006c) goes on to list the criteria that may shape how work is allocated across a system.

- Actor competencies
- Access to information or means for action
- Level of coordination
- Workload
- Safety and reliability
- Regulatory compliance
- Enjoyments
- Availability.

Worker Competencies Analysis

The final phase of the CWA framework, Worker Competencies Analysis, involves identifying the competencies required by agents for performing the activity within the system under analysis. According to Kilgore and St-Cyr (2006), the goal of Worker Competencies Analysis is to identify psychological constraints applicable to systems design. As the final phase of the CWA framework, WCA inherits all requirements identified through the four previous phases. It is important to note that the output of WCA is not a finished design. Instead, the entire CWA process feeds constraints for developing information requirements used in subsequent interface design activities. Kilgore and St-Cyr (2006) go on to say that several theories and models are relevant to identifying the implications of human characteristics in system design (e.g. manual control models, sampling theory, signal detection theory, etc.); however, each of these models focuses on specific, narrow psychological traits. Instead, an integrated model is needed to aid designers in deriving practical implications for system interfaces. Vicente (1999) proposes the use of the SRK taxonomy to address this need, Figure 2.12 shows each of the three types of behaviour along with their cues.

According to Vicente (1999), Skill-Based Behaviour (SBB) is performed without conscious attention. SBB typically consists of anticipated actions and involves direct coupling with the environment. SBB is mainly dependant on automatic responses and neuromuscular control. Rule-Based Behaviour (RBB) is based on a set of stored rules (if-then) that can be learned from experience or from protocol. Individual goals are not considered, the user is merely reacting to an

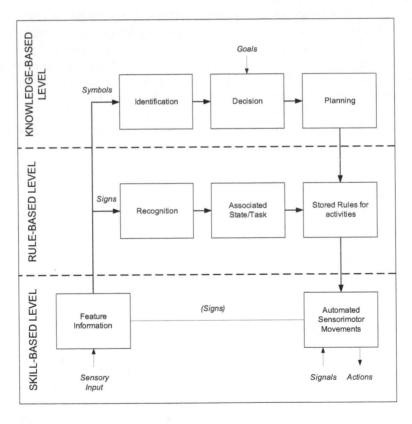

Figure 2.12   The SRK taxonomy of human performance categories (reproduced from Vicente 1999)

anticipated event using familiar perceptual cues. Unlike SBB, users can verbalise their thoughts, as the process is cognitive. When decisions are made that explicitly consider the purpose or goal of the system the behaviour can be considered to be Knowledge Based Behaviour (KBB). KBB is slow, serial and effortful because it requires conscious, focal attention. According to Vicente (1999), interfaces for complex sociotechnical systems should encourage SBB and RBB, while allowing for operators' seamless transition to KBB during problem solving.

Tasks rarely fall into one class of behaviour; the three levels interact. Novice users may find that they start to reason at a knowledge-based level considering each task they carry out. Through experience, this knowledge-based behaviour can turn to rule-based behaviours as familiar perceptual cues are recognised. According to Vicente (1999) the outcome of a worker competencies analysis should be a number of detailed implications for interface design and training, and using knowledge of the signs, signals and symbol distinctions, information can be presented in a form that will make it more likely that particular levels of cognitive control will be activated. As with the Strategies and Social Organisation and Co-operation Analysis components, the Worker Competencies Analysis component has received only minimal attention to date.

To support this phase, Kilgore and St-Cyr (2006) developed an SRK inventory (see Figure 2.13), the conventional table used to capture SRK information (shown in white) is accompanied by additional boxes to capture information about the specific activity taking place (shown in grey). The cells within the first two columns describe the information processing activities and resultant knowledge states. The population and organisation of the rows within these columns can be informed directly by decision-ladders generated during ConTA activities; thus, creating a direct link between the outputs of the Worker Competencies Analysis and previous CWA phases. Kilgore and St-Cyr (2006) argue that if there are sections of the table that are incomplete then the interface is not performing as well as it could be. An empty cell in the Knowledge column indicates the interface may not provide an operator with sufficient resources to engage in effective troubleshooting, or problem solving, for a particular information-processing step. Kilgore and St-Cyr (2006) go on to point out that, completed cells in the SRK Inventory can also be used to generate profiles of competencies that operators must possess to perform control tasks adequately. In this manner, the SRK Inventory can be used to inform worker selection and training. Figure 2.13 shows an example of the SRK Inventory completed by Kilgore and St-Cyr (2006) during an analysis of an air traffic control simulation. The scenario captured is from a ConTA activity designated as 'Rerouting'. An operator detects multiple aircraft with potentially conflicting flight paths and then determines and relays flight instructions to these aircrafts to avoid collision. The diagram proposed by Kilgore and St-Cyr (2006) has been extended to show graphically the processing steps on mini decision-ladders.

Information processing step	Resultant state of knowledge	Skill-Based Behaviour	Rule-Based Behaviour	Knowledge-Based Behaviour
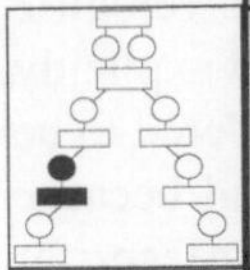 Scan for aircraft presence in area of responsibility	Whether multiple aircrafts are within area of responsibility	Monitoring of time-based spatial representation of aircraft in area of responsibility	Perceive explicit indication multiple aircrafts are currently within area of responsibility	Reason, based on proposed flight plans, that multiple aircrafts may be present in area of responsibility within similar time frames
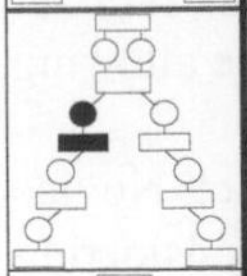 Determine future flight vector for each aircraft	Whether multiple aircraft within area of responsibility have intersecting flight paths	Perceive headings of related aircraft as convergent, divergent, or parallel	Use heuristics to determine whether flight paths are intersecting (e.g. flight paths A-B and C-D are convergent paths; flight paths A-C and B-D are divergent)	Reason, based on geospatial knowledge of to/from points for each flight, that aircraft are on convergent or divergent paths
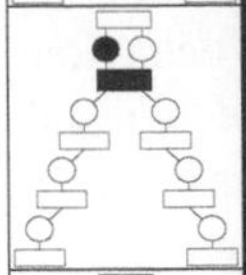 Predict future, time-based location states for aircraft on convergent paths	Whether converging aircraft will arrive at point of convergence within a similar time frame	Perceive time-to-collision (tau) of each aircraft with the convergence point, based on spatial representations of heading and speed	Use heuristics to estimate whether aircraft will arrive at convergence point within a similar time frame (e.g. if distance A = distance B AND airspeed A = airspeed B, aircraft will arrive at approximately same time)	Calculate, using airspeed, heading, and location of each aircraft, the time at which each aircraft will arrive at the convergence point
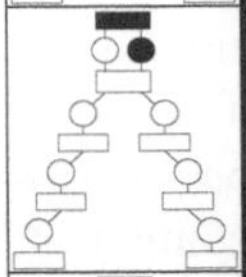 Determine the criticality of a pending convergence	Whether future distances between converging aircraft will constitute a 'loss of separation' event	Perceive whether the zones of safe travel surrounding each aircraft will overlap at or near their closest point	Use heuristics to determine proximity as being greater or less than the minimum required envelope of separation	Calculate distance between each aircraft at their closest future states and compare with the minimum value of separation required for safe travel.
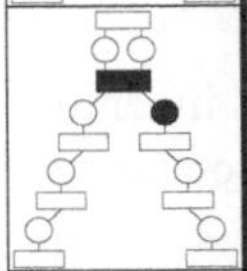 Choose to modify aircraft flight path(s) to address future problem	Which aircraft flight path(s) must be modified to eliminate potential 'loss of separation' event	Directly perceive that one or more aircraft must be redirected	Apply doctrine: (e.g. If loss of separation will occur, MUST reroute one or more aircraft)	Reason from knowledge of proposed flight paths, current locations, and expected future behaviour that aircraft must be rerouted
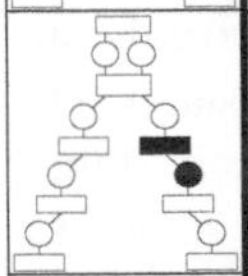 Select specific method for accomplishing rerouting of aircraft	Air Traffic Controller's awareness of new flight path(s)	Respond automatically to perception of loss of separation by directly manipulating a representation of aircraft flight path(s)	Classify loss of separation within a set of generalized scenarios and select appropriate stereotypical control rule	Develop new, optimized flight path(s) based on weighted criteria including urgency, flight priority, passenger convenience, efficiency etc.
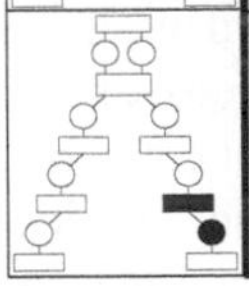 Convey flight modifications to aircraft for execution	Aircraft's awareness of new flight path(s)	Direct, simultaneous interaction with communication equipment through control interface through input of rerouting information (step 11)	Apply stereotypical control rules to select method/sequencing for conveying proposed flight path (e.g. if one aircraft involved, contact that aircraft; if two, contact both)	Reason using knowledge of aircraft systems, priorities, urgency, etc., the best means and order for contacting each aircraft to convey proposed flight path(s)

Figure 2.13   SRK Inventory (adapted from Kilgore and St-Cyr 2006)

Chapter Summary

This chapter has identified the need for a framework capable of modelling complex sociotechnical systems, the chapter started by exploring the problem of complexity and 'command and control'. It then introduces CWA as a possible method for addressing this complexity; thus far, the framework has been described in a theoretic manner, as with many things 'the proof of the pudding is in the eating', the subsequent chapters apply the methodology to explain it and further test its suitability.

CWA has particular applicability to product designers as it allows the modelling of 'first of a kind' systems. The Work Domain Analysis phase in particular, leads the designer to focus in a structured way on the reason for developing the system/ product. The framework encourages the designer to work based on the constraints of the system, whilst focusing on the key criteria by which the system/product will be evaluated. The process of considering the functions in abstract terms allows for creative thinking and problem solving. This encourages the designer to consider the need they are addressing, rather than jumping straight in to solving the problem. The different phases and tools within the CWA framework can be applied throughout the design life cycle. CWA applications have previously been used for purposes ranging from the design of novel systems to the analysis of operational systems. This chapter has highlighted the differences between formative approaches, such as CWA, and normative approaches; it is contended that CWA is better equipped to cope with the levels of flexibility within complex sociotechnical systems. According to Naikar (in press) the highly dynamic nature of network-centric warfare means that the goals and work requirements of participating systems, and the humans operating those systems, are frequently changing – often in ways that were not, or even cannot be, anticipated by engineering designers or by professional, trained workers. Furthermore, rapid developments in the information and communication technologies that enable network-centric warfare means that the nature of human work is continually evolving. Standard techniques for work or task analysis are not well suited for modelling activity in network-centric operations. Instead, a formative approach is necessary and CWA provides such an approach. Lintern et al (2004) state that the constraint-based approach of Cognitive Work Analysis is one that can cope comfortably with the scale-up problem. Indeed, the potential benefits from a Cognitive Work Analysis grow as systems become larger, more technologically sophisticated, and more complex.

CWA is not without its limitations, its formative nature, the great strength that separates it from other methods, can also be perceived as it weakness; CWA is notoriously non-prescriptive which has implications for its ability to be taught and understood. According to Fidel and Peijtersen (2005), because CWA investigates information behaviour in context; individual studies create results that are only valid for the design of information systems in the context investigated, rather than for the design of general information systems. Results from a variety of studies; however, can be combined together and generalised to inform the design of other information systems. Because of the flexibility in the application of the tool, the outputs of the analysis are likely to be very different for different analysts, the decisions made regarding the system boundaries are key to the output. The application of CWA typically requires substantial resources to produce large and complex outputs. From the available literature it is clear that CWA is more commonly applied in industries that can support and finance the time it takes analysts to learn and apply the technique; such as; aviation, power generation, medical and military domains. Much of the existing literature has a strong focus on WDA. The social-organisation and co-operation analysis, and worker competencies analysis components of the CWA framework, have received only limited attention; consequently, there is very little published literature regarding the application of these phases. The analysis requirements dictate which of the CWA phases are applied in an analysis effort. CWA is not intended to replace all other human factors techniques, like any tool it is best used for its appropriate purpose. The approach forms a good basis for starting an analysis; it is complimented by other more prescriptive methods that can provide additional detail. The remainder of this book will investigate how well CWA supports the analysis, development and evaluation of complex sociotechnical systems; it will also suggest refinements to the framework to enhance this capability.

Chapter 3
Interaction Design

Chapter Introduction

This book would not be complete without some discussion of fundamental design principles. The aim of this chapter is not to completely critique the literature surrounding interaction design, we fear this topic would take an entire book; instead, the aim is to briefly explain some basic guidelines that should be considered in the design of any product or system.

Art or Science?

Both Saffer (2007) and Burns and Hajdukiewicz (2004) are clear to communicate that they consider design to be an 'art'. By its very nature design is about the conception and development of new products; in creative and exploratory ways. Design is often heavily context dependant, with products developed for a particular scenario and time. No matter how well defined or constructed a methodology is, design can never truly be turned into a science (i.e. comprising the immutable laws of behaviour). While both Saffer (2007) and Burns and Hajdukiewicz (2004) agree with this, Burns and Hajdukiewicz (2004) describe EID as being in some way akin to painting by numbers, they comment that; while great design cannot be taught, the underlying principles can. Tufte (1983) on the other hand is rather sceptical of design principles:

> Design is choice. The theory of the visual display of quantitative information consists of principles that generate design options and that guide choices among options. The principle should not be applied rigidly or in a peevish spirit; they are not logically or mathematically certain; and it is better to violate any principle than to place graceless or inelegant marks on paper. Most principles of design should be greeted with some scepticism, for word authority can dominate our vision, and we may come to see only through the lenses of word authority rather than with our own eyes. (Tufte 1983)

The issue of where design sits on the 'art-science continuum' seems extremely contentious. There seems to be general agreement that good design can be informed by analysis; however, the correlation between good analysis and good design appears to be poor, it seems that great designers manage to consider a multitude of factors that seem impossible to capture and replicate. It would appear that, what is needed is a structured approach to design that is able to guide development, whilst retaining the necessary flexibility so as not to overly constraining the creativity, or art, of design. Arguably, there is less of a requirement for a structure in the design of simple artefacts; in these cases, the constraints governing the products are frequently clear and well understood, the use of basic principles combined with a designer's creativity is often enough. When we turn to the design of interaction in large, complex system, however, the complexity rises significantly and, habitually, the constraints are, less well understood. To create effective designs in these domains the factors that need to be considered are, more often than not, too numerous to consider without some method of explicitly structuring them. As Don Norman (2007) points out:

> As we move towards the design of intelligent machines, rigour is absolutely essential. It can't be the cold, objective rigor of the engineer, for this focuses only on what can be measured as opposed to what is important. We need a new approach, one that combines the precision and rigor of business and engineering, the understanding of social interactions, and the aesthetics of the arts. (Norman, 2007, p.173)

What Norman makes clear is that a symbiotic relationship is required between science and art. In such complex systems, it seems highly unlikely that great interaction design can be achieved without consideration of the two. This chapter focuses, primarily, on the 'art' of design and the basic principles that support it. The remainder of the book then discusses the 'science' based approach that supports it; and how the two can be integrated.

Interface Design or Interaction Design?

Before starting to describe the basic principles of design it is worthwhile defining some of the terms used in this book. The concordance between Saffer's (2007) book on 'Interaction Design' and Burns and Hajdukiewicz's (2004) book on Ecological Interface Design (EID) seemed particularly apparent. Despite the books' differing target audiences (Saffer's book which it is assumed is aimed at those from a design background, and Burns and Hajdukiewicz's book which is aimed more at those from a human factors or engineering background). The most obvious difference between the two approaches is that one (Saffer's) refers to interaction design and the other refers to interface design. Saffer (2007) defines interaction design as follows:

> Interaction design is the art of facilitating interactions between humans through products and services. It is also, to a lesser extent, about the interactions between humans and those products that have some kind of 'awareness' – that is, products with microprocessors that are able to sense and respond to humans. (Saffer, 2007, p.4)

Saffer (2007) points out that, as it is often invisible, it is not always easy to define exactly what interaction design is; his discussion of the Windows and

Mac 'operating systems' (OS) serve as good examples; whilst the two operating systems can appear visually to be very similar, the 'feel' or user experience can be completely different. Essentially interaction design is a broader term that encompasses interface design.

Interface Design Principles

The modern world is flooded with man-made interfaces, they exist in almost every product we use; cash machines, cars, computers, televisions, phones, right down to products as simple as toothbrushes. A design interface is effectively made up of one or more pieces of information that are communicated to the user (and in most cases can be manipulated). The layout and complexity of the interface will be influenced by both the complexity of the task, the display and the design and arrangement of the nested pieces of information.

Why Should it be Shown?

Ware (2000) defines visualisation as 'a graphical representation of data or concepts,' which is either an 'internal construct of the mind' or an 'external artefact supporting decision making'. In other words, visualisations assist humans with data analysis by representing information visually. According to Troy and Möller (2004) visualisations can provide cognitive support through a number of mechanisms, these mechanisms can exploit advantages of human perception, such as parallel visual processing, and compensate for cognitive deficiencies, such as limited working memory (see Table 3.1).

What Should be Shown?

Essentially, interface design is about selection; Troy and Möller (2004) point out, that data sets are often very large; however, only a limited number of graphic items with limited resolution can be concurrently displayed on a display. Thus, displaying more items often means displaying less detail about each item. If all items are displayed, few details can be read, but if only a few items are shown, the user can lose track of their global location. This requires users to retain large amounts of either detail or context information in working memory, producing an extra cognitive load that may affect performance. Researchers such as Kasik (2002) have investigated the effects of increasing the screen size; however, this can only be done to a limited extent before the screen becomes too large to scan. Tufte, in his groundbreaking trilogy of texts (Tufte 1983; 1990; 1997), discusses many issues associated with effective graphical communication; he places a great emphasis on visual clarity and a move away from what he terms 'chartjunk'; 'chartjunk' refers to unnecessary visual clutter or information on the screen. Tufte favours interfaces with the minimal amount of 'ink' and he strongly discourages superfluous detail

Table 3.1   How information visualisation amplifies cognition (Troy and Möller, 2004, adapted from Card et al, 2004 pp.2)

Method	Description
Increased Resources	
Parallel processing	Parallel processing by the visual system can increase the band width of information extraction from the data.
Offload work to the perceptual system	With an appropriate visualisation, some tasks can be done using simple perceptual clues.
External memory	Visualisations are external data representations that can reduce demand on human memory.
Increased storage and accessibility	Visualisations can store large amounts of information in an easily accessible form.
Reduced Search	
Grouping	Visualisations can group related information for easy search and access.
High data density	Visualisations can represent a large quantity of data in a small space.
Structure	Imposing structure on data and tasks can reduce task complexity.
Enhanced Recognition	
Recognition instead of recall	Recognising information presented visually can be easier than recalling information
Abstraction and aggregation	Selective omissions and aggregation of data can allow higher level patterns to be recognised
Perceptual monitoring	Using pre-attentive visual characteristics allows monitoring of a large number of potential events
Manipulable medium	Visualisations can allow interactive exploration through manipulation of parameter values.
Organisation	Manipulating the structural organisation of data can allow different patterns to be recognised

such as cross hatching and unnecessary borders. Tufte also has a strong respect for the statistical validity of the interface or graph, believing that clear concise descriptions can prevent the user from misinterpreting the information or placing a biased emphasis on a particular part.

Unsurprisingly, the way a display is used also informs the way the data should be represented; in his seminal paper, Grether (1949) contends that displays [in aviation] are read in three ways depending on the purpose of the reader. He goes on to categorise them as follows:

1. Quantitative reading – for a numerical value of an indication. Quantitative displays show exact information. Information is presented in a way so that exact static numerical values can be understood. Examples of this would be clocks and watches where the exact time is required, or thermometers where the exact current temperature is required.
2. Qualitative reading – for the direction and approximate magnitude of a deviation from the null, normal or desired indication. Qualitative displays give information about particular states, in context, information here is presented in relation to other information, to show rates of change, current state, etc.
3. Check reading – for the assurance of a null, normal or desired indication. Check-reading displays are a specific type of qualitative display these can be used to determine whether the value of a continuously changing variable is normal, or within an acceptably normal range, for example, car fuel gauges and tyre pressure gauges. Check-reading displays should have clearly distinguishable characteristics to identify the neutral or normal satisfactory condition, or the undesirable condition; perhaps green marking for an 'OK' level and red for 'out-of-limits'.

Although Grether (1949) describes these categories for the purpose of display evaluation, it is reasonable to use Grether's evaluation categories as a method for informing the design requirements of a display.

The Nuts and Bolts of Design

The process of designing a display can often be informed by considering the optimal representation for the different data sets. According to Burns and Hajdukiewicz (2004), there are three different ways of referring to an individual piece of data: propositionally, iconically, or analogically.

Propositional Forms

Propositional forms are abstract representations of the data; they include written and spoken language. According to Burns and Hajdukiewicz (2004) propositional forms like language are, to a certain extent, arbitrary and need to be learnt. For expert users, propositional forms are natural and automatic; however, for novice users they are unintuitive and require effortful translation. Propositional forms are suited to situations with expert users where new users have the access to comprehensive training. Much of military symbology can be considered propositional; these would include the symbology used by military planners detailed in MIL-STD-2525B (DOD, 1999). These representations rarely have any direct relationship to the objects they represent (see Figure 3.1).

One of the simplest forms of display is the digital numerical display (see Figure 3.2). This display communicates exact information on the state of the system in a propositional form. This kind of display is ideally suited to the task of matching the car's speed with a speed restriction within the environment.

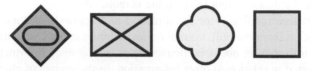

Figure 3.1   Examples of propositional forms (from left to right; enemy armoured unit, friendly infantry unit, unknown unit, neutral unit)

Figure 3.2   Digital display

Iconic Forms

Iconic forms are metaphoric representations of the data; they contain a semantic link to the source of their representation. Iconic forms are used throughout the car instrument panel to represent indicators, hazards, ice, lights, etc (see Figure 3.3). These iconic representations are ideally suited as they transcend language barriers. The success of iconic representations is dependant on their design creating a semantic link to the object in the world they represent.

Iconic signs do not necessarily always map the real world exactly; a good example of this is the departures and arrivals signs used in airports (see Figure 3.4). The image on the left shows the sign for departures; this closely represents what a plane looks like taking off. The image on the right is the sign for arrivals; those who have ever seen a plane land will know that the plane approaches the runway with its nose up, like in the departures sign, however, to distinguish the two the image on the right has been adopted.

Figure 3.3   Iconic representation for seat belt, oil and battery

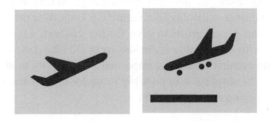

Figure 3.4   Iconic signs for departures (left) and arrivals (right)

Figure 3.5 shows the iconic representation for no smoking, in this case, this symbol relies partly on a propositional form, the red circle with a line across it is partly iconic being red and crossing something out; however, it is propositional in the way that its understanding of don't do what is inside that may need to be learnt.

Analogical Forms

According to Burns and Hajdukiewicz (2004), analogical forms capture some sort of constraint in the environment, which map that constraint on to a visual

Figure 3.5   Iconic representation for no smoking (in colour the circle with a line through it is red)

form. An example of this is a map; the map supplies a scale representation of the environment.

Choosing the Representation Type

When conducting the task of matching the speed of a vehicle to the legal speed limit the only useful contextual information is the current vehicle speed and the current speed limit. There is little or no need for the driver to understand the maximum or minimum possible speed of the vehicle while conducting the task of speed matching. Context becomes far more important where maximum and minimum levels are concerned, for example, on an automotive rev-counter. Arguably for the task of speed matching, a digital read out as shown in Figure 3.2 is more appropriate than an analogue between limits gauge. Where context is required, such as a car rev counter, a display like Figure 3.6 is more appropriate.

Target states can be added in to give further context. Figure 3.7 shows an analogue display with undesirable states shown in red (the dark shaded area).

Sometimes the context of time is important in a display to show a trend. By adding history to the display, it is possible to understand if the current value is increasing or decreasing (see Figure 3.8).

Figure 3.6   Examples of analogue display with context

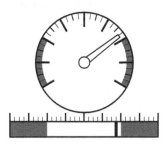

Figure 3.7   Examples of analogue display with context and alarm states

Further context can be added to Figure 3.8 adding on limits (see Figure 3.9)

When more than one set of data needs to be compared, additional information needs to be displayed. Typically, this can be displayed on bar charts or histograms, radar plots (see Figure 3.10) also allow a number of variables to be readily compared. The use of polygons is seen as being a superior form of display, not only for multiples of integrated variables but because they are processed

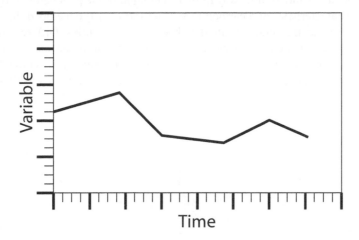

Figure 3.8   Analogue display with history

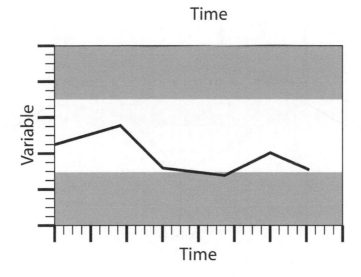

Figure 3.9   Analogue display with history and limits

holistically. Barnett and Wickens (1988) also found that polygons were better than conventional displays in fault diagnosis. Munson and Horst (1986) supported this by using polygons to test normal/abnormal states increasing the number of vertices analogous to abnormalities. They found that reaction times decreased when more vertices were present, concluding that polygons were processed in parallel, and equivalent to holistically.

A more novel and abstract way of representing multiple parameters is by using different characteristics of a cartoon face. Chernoff (1973) developed this idea as a way of representing multivariate data back in the seventies. Different features of the face were made to correspond with data points in dimensional space, e.g. length of nose or curvature of mouth. This type of representation was thought to be an ideal way for people to be able to visualise different data points as they could

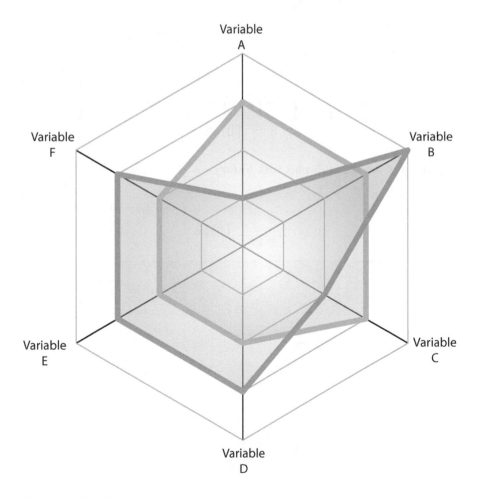

Figure 3.10   Radar plots

easily assimilate information from a well-known graphical representation. It was also easier for them to identify irregularities in the data. In his initial experiments, Chernoff (1973) used 18 different parameters (facial features) to examine different data measurements of fossils, and found this type of representation useful in cluster analysis and detecting changes in time series. Another advantage of using the face as a graphical representation for multiple parameters is that information identification does not appear to become more difficult as the number of variables increases. In fact, it is reported that as the number of parameters containing useful information increases, so the information transmitted appears to be richer in content. Whilst these displays perform well in the task of change detection, it is arguable that these representations are propositional. The relationships to the underlying variables are arbitrary and need to be learned. Examples of the original Chernoff's faces are given below in Figure 3.11.

Developing Displays

Based upon an understanding of the type of data to be represented it is possible to limit down the type of display to a small subset. The choice can be informed by a number of questions:

Single Displays

- Is it binary (two states) / multi-state / analogue?
- Is it between limits?
- Are the limits fixed or variable?
- Is it critical to monitor?

Figure 3.11 Example of Chernoff's faces showing either five or seven variables for identifying measurements of fossil specimens

- Does trend need to be tracked?
- Does rate of change need to be tracked?

Multiple Values

- Is the value additive?
- Do values need to be considered in context to each other?

The following flow chart has been completed to aid in the selection of displays; the decisions for the displays are based on Burns and Hajdukiewicz (2004).

Grouping Displays

Displays can be arranged based on a number of criteria. The choice of these criteria is dependant on the system represented.

Spatial Importance

Spatially arranged displays are used where the relative location of data sets is important. An example of this could include a battlefield display where the relative location of units in the field is of paramount importance.

Functional Importance

Functionally arranged displays may be used to monitor manufacturing processes. The display may be constructed to represent the flow of the product through the manufacturing process to supply information at each stage. The diagram may not exactly match the physical layout of the factory; it may represent more of a schematic.

Hierarchical Importance

Hierarchically arranged displays can be used to show data in some kind of hierarchy, for example, with the data related to the most senior objects at the top and the data relating to the least senior objects at the bottom.

Critical Importance

Critically arranged displays represent data based on its importance. In this kind of arrangement important data would take a prominent position and be likely to be larger, and therefore more salient, than less critical data.

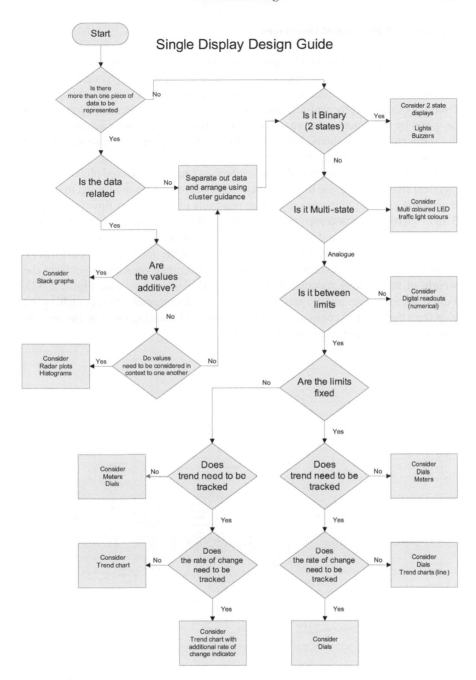

Figure 3.12   Single display selection flow chart

Physical and Functional Interfaces

As the titles suggest, functional interfaces concentrate on the overall function of the object, whereas, physical interfaces focus on representing the physical elements within the system. A functional interface focuses higher up the Abstraction Hierarchy than the physical interface. A simple example of these differences can be given by the functional purpose of heating food in a domestic setting. Two of the more common methods, within a domestic environment, are a standard oven (gas or electric) and a microwave oven.

In most cases, a conventional oven's interface can be considered physical. Here physical changes are made to mechanical components; knobs are turned to change thermostat cut-off points, or vary valve positions to change gas flow (see Figure 3.13). Microwaves; however, work at a far more functional level. Here settings are chosen (cook, warm, defrost) and a time is attributed to these functions (often read directly from a packet or from guidelines on the face of the microwave). More advanced microwave models are able to determine cooking times from the weight of the product, calculating the time and power levels (see Figure 3.14). Functional interfaces usually rely on some level of automation, thus, detaching the user from the physical components of the system and focusing them on the outputs and purposes they are attempting to achieve.

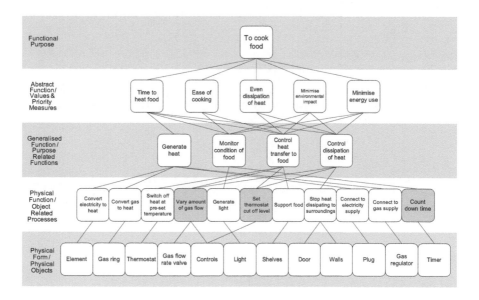

Figure 3.13   Abstraction Hierarchy for a conventional oven (highlighted boxes show how the system is manipulated in this case by making changes to object related processes)

The interface design of a product has the ability to influence how it is interpreted and used, as either a functional or physical interface. Whilst microwaves are considered functional objects, it is still possible to think of them in physical terms. The differences between physical and functional interfaces can be further explored by comparing two microwaves. An example of a more physical interface can be seen on the left hand side in Figure 3.15. The user selects the amount of energy and the length of time the food will be exposed to this energy. In the example on the right hand side in Figure 3.15, the user still has the option to operate the microwave in a physical way; however, the option exists for the user to use the interface in a far more functional way. Tailored cooking programs exist for reheating a number of common foods such as curry, Chinese food, pasta, and for cooking a number of common foods; jacket potatoes, fresh vegetables, and fresh fish. Rather than manually adjusting the cooking time and the energy, the type of food and weight is entered into the microwave.

Metaphoric References

Simply by looking at mechanical objects with exposed components (scissors, tin openers, corkscrews, bicycles, etc) it is very easy to develop a mental model of

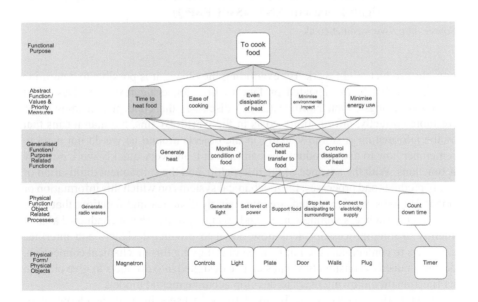

Figure 3.14   Abstraction Hierarchy for a microwave oven (highlighted boxes show how the system is manipulated in this case by making changes much higher up the hierarchy)

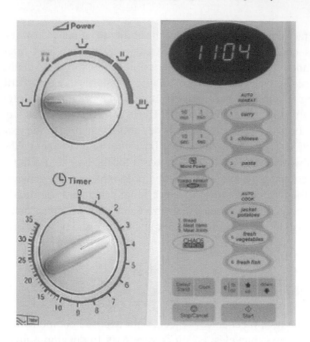

Figure 3.15   Physical (left) and functional (right) microwave interfaces (Example shown left: Proline SM18 WHITE; Example shown right: Panasonic NNT545W1 FBPQ)

Source: http://www.comet.co.uk

how the product might work. By examining the object, it is possible to determine its affordances. In the updated preface to his pioneering text 'The Design of Everyday Things', Don Norman (2002) discusses the concept that knowledge (or more correctly, information) exists in the world. By this, Norman means that as the information is stored visually in the objects around us we do not need to remember how to use a product; we can look at it and very quickly work out how it functions. With the advance of the microchip, and the subsequent digitisation of many products and processes, the mechanical systems on which this information or 'mental models' were extracted, are increasingly disappearing; functions that were once visible such as spinning wheels and swinging arms are not conducted almost silently in microprocessors. It is now down to product designers and interface designers to contemplate the implication of removing this mechanical connection, designers need to consider how cues can be designed into products to give clues to its functionality allowing users to develop accurate mental models.

Metaphoric references can be used to assist users in their development of mental models. Accurate mental models can have a significant impact on the time taken to learn how to use the product; requirements for formalised training; user satisfaction; and error rates. One common method of incorporating metaphoric references into a product is to design in a graphical representation of the previous, real world, situations that work in similar, predictable ways. This method has been used many times in the past. The 'Windows' graphical user interface (GUI) includes metaphoric relationships for files, folders and wastepaper bins. This interface allows the user to drag a file from one folder to another physically or to delete a file by dragging it from the folder to the wastepaper bin (which can be accessed as long as it has not been emptied). This kind of interaction is often termed 'direct manipulation' (Shneiderman, 1983, 1998), whereas, when using indirect manipulation more abstract commands or menus are selected, direct manipulation mimics actions in the physical worlds such as dragging, rotating and resizing. Another example of metaphoric references is used by some mobile phone manufacturers; when a text message is received, it can be seen to fly in as an envelope and open up. The reverse happens on the sending of a message where the message is folded and placed in an envelope and sent out. These analogies are metaphoric as they link back to physical representations. It is clear that there is no need for an envelope when sending the digital message; however, from the physical world users have developed a semantic link between the envelope and the sending of information. In certain cases, theses metaphoric references become so established they outlive the products they once represented. The icon used to save files in many computer applications was designed to resemble the 3.5-inch floppy disk. Whilst the icon has stood the test of time, it is now extremely uncommon to see the physical disks used. Norman (2002) advocates the use of natural mapping as a way of taking physical analogies for immediate understanding; he gives the example of moving an object up by moving the controls up. Norman goes on to discuss the difficulty in mapping subjective dimensions (such as taste, colour and location) where a plot of less and more can be much harder to map.

According to Norman (2004), 'if you can't understand a product you can't use it – at least not very well'. Whilst it is possible to remember a set procedure for operating a product, each of these procedures needs to be learnt and memorised. Without some kind of cue these are often forgotten; according to Norman (2004) 'learn once remember forever' ought to be the design mantra, one way of achieving this is to store information in the environment rather than in the head.

Figure 3.16 shows graphically the theory of conceptual models. The designer builds a conceptual model of how the product functions and imparts this information into the interface (uses the interface to communicate their model visually). The user then looks at the interface, extracts the information and builds a conceptual model of the product. Other than through formalised training or instruction manuals the interface is the only way the designer can communicate their conceptual model. The success of communication is heavily influenced by the skill of the designer. According to Norman (2004), the product's success in correctly transferring an accurate mental model can be significantly increased by an iterative design process involving the development and evaluation of a series of 'mock-ups' in the early stages of the design process. The evaluation of these interfaces should take place with end-users and should be observed by professional

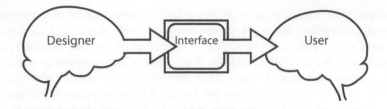

Figure 3.16   Conceptual models (adapted from Norman 2004)

observers. Whilst this approach is valued, care should be taken when applying it to large sociotechnical systems; in many cases, it is unsuitable to rely on users to supply accurate subjective opinions on how to improve the product.

When Mental Models are Incorrect

Norman (2004) discusses the implications of having poor mental models, and the resultant issues with trust that arise with these products. Much of the problem stems from the user not knowing what to expect with all of the operations invisible and hidden from view. Norman (2004) states that without any way of users understanding how devices operate or what actions are required, users can develop feelings of loss of control and disappointment. Norman (2002) gives a good example of how mental models can often be incorrect using a domestic oven as an example. When observing consumers preheating domestic electric ovens to a set temperature, it is quite common for the user to turn the oven on to full temperature in the, misguided, belief that it will get hotter quicker, users then reduce the temperature down to the desired temperature. This incorrect assumption is due to the mental model formed by the user. Norman (2004) points out, that there are two commonly held folk theories about thermostats, the timer theory and the valve theory. The timer theory states that the thermostat controls the relative time the device is on; half way means the device is on half of the time. The valve theory is that the thermostat affects the amount of heat entering the oven. The correct model is that the device can only have two states, fully on or fully off and the thermostat is a switch activating these two states based on the temperature of the oven. In the case of the thermostat, there is no reference to how the thermostat works; the user is therefore required to build the model on past similar experiences.

Vicente (1997) expresses his concern over displays that are developed from user mental models. Vicente points out that if the model is incorrect or incomplete then the resulting interface will in turn be incomplete or incorrect. Vicente also acknowledges the significant differences that exist between mental models generated by different users. Vicente offers the following anecdote to describe this problem:

One nuclear power plant vendor chose an exceptionally skilled operator to participate in the design of a new control room. This practice is consistent with existing theories of 'participatory design' or 'user-centred design', the goal being to get user input into the design process. This is a laudable goal and represents a distinct improvement over traditional design practices, which essentially leave the user out of the design process. Nevertheless, the moral of this anecdote is that it is possible to take this user-centred approach too far. When the new control room design was shown to other operators, the designers quickly realized that the design process they had adopted was faulty. The first valuable lesson they learned was that other operators did not think of the plant in the same way as the exceptional operator who was part of the design team. In fact, no two operators seemed to have the same mental model of the plant. This finding caused the designers to reflect more deeply upon the relationship between operators' mental models and the engineering laws and principles governing plant behaviour. This reflective exercise revealed another valuable lesson, namely that all of the operator's mental models were limited in the sense that they contained misconceptions, omissions, or both. As a result, the control room in question had to be redesigned to reflect better the way in which the plant actually worked. Needless to say, this was a costly and time-consuming process. Vicente (1997)

Whereas, approaches that are more traditional tend to put a great deal of emphasis on analysing human characteristics (e.g. Wickens, 1992), the ecological approach (e.g. Rasmussen, Pejtersen, and Goodstein, 1994) puts much more emphasis on analysing the interaction between people and their environment. The ecological approach moves away from focusing on the individual, it begins by analysing the environment before analysing what people are doing, how they are doing it, or what they know. This ecological approach encourages users to develop mental models that are based on the constraints of the actual system rather than based on a more normative description captured from an expert user.

Standardise Approaches

Another way to lessen the burden on memory is to adopt standard approaches. These approaches apply to a 'learn once apply many times' mantra. Although these standards may be abstract, once learned they may be adopted into a language much in the same way as propositional forms (see Chapter 3). For the interaction with computers there are a set of established and frequently used standards; examples include keyboard shortcuts.

Copy something	Ctrl-C
Paste something	Ctrl-V
Undo the last action	Ctrl-Z
Select an object	Left mouse click
Access an object related menu	Right mouse click
Access additional info	Mouse over and wait for text to pop up

Help   F1
Reapply the same operation   F4

The question arises 'when do we follow standard?' Perhaps the decision is made for us by Alan Cooper's axiom 'obey standards unless there is a truly superior alternative'.

What are We Designing? – A Case Study for Military Planning

The following case study has been included to illustrate how some of the discussed principles can be applied. The example of digital battlefield planning is used; digital battlefield planning is a key concern for any military organisation concerned with distributed planning and the real time tracking of friendly and hostile units. Ever increasing sensing and networking capabilities have made it possible for an enormous amount of information to be available to the command team. The design and choice of representation of this information has massive implications for the performance of the system. One particularly rich source of metaphors for this process is the physical environment in which the planning previously took place (see Figure 3.17). This environment consisted or a large wooden table with a large map spread across it. Individual units and battalions were represented on this map with 'stickies' (small pieces of acetate with symbology taken from MIL-STD-2525B printed upon them; see examples in Figure 3.1). Sheets of acetate are spread over the map; additional graphical information can be added by drawing directly onto the acetate. 'Post-it notes' can be stuck on the map to add additional textual information. Different situations can be shown on different overlays. Due to their transparent nature, it is possible to view two overlays at the same time.

To develop new and better ways of doing things, it is fundamentally important that new systems are not constrained by previous ways of doing things; however, at the same time many of these existing analogue processes have developed over long periods of time; they have been refined based upon experience. A careful balance needs to be made between embracing new technologies whilst still retaining the applicable references to older established systems; to coin a phrase 'we don't want to throw out the baby with the bath water'. Observations were made of the paper-based approach in a training exercise, an early evaluation of a digital prototype was also observed. From a week observing users attempting to use the new tool in the planning process, a number of opportunities were observed. Among these observations, two specific scenarios stand as clear examples where metaphoric relationships were requested by the user indirectly.

Scenario 1

In the digital system the annotation of a 'user defined overlay' (UDO) is synonymous with the marking up of a sheet of acetate. In the observed computer digital system

Figure 3.17 Traditional bird-table

the UDO is created using indirect manipulation through drop-down menus (From the drop down menu, select Situation > Create/Modify > User defined overlays). This opens a new window at the left hand side of the current map window. Symbols are then added to the map (via a, rather unique, sequence of left and right mouse button clicks) to create the UDO. Once the objects are added an additional process is required to associate them with the UDO; they need to be copied and pasted onto the UDO (by left clicking the symbol and then right clicking on the UDO bar on the LHS of the screen). One of the users pointed out:

> This is unintuitive as we already created the UDO why do we need the last step? In the analogous physical system the UDO would be created by rolling on a new sheet of acetate any subsequent annotation or 'stickies' would go directly on to this acetate or UDO.

This anecdote describing an early prototype provides one example of where the designers did not get it quite right. The product appears to have been developed from an engineering approach that has not taken enough lead from the environment that it models, or from the activities required within it (neglecting to build upon any mental models). There is a introduction of strange new practices with the mouse that move away from standardised approaches which appear to have little benefit and caused significant annoyance. The choice of indirect manipulation also causes problems making the described process long and drawn out with little cue in the environment to prompt the user, this places a requirement on the user to learn the process. As Tesler's law tells us, we cannot hide the complexity in the system, but we can get the computer to do some of it for us, many of the operations observed could be done automatically by the computer and replaced with a far more intuitive direct manipulation 'drag and drop' interaction. A consideration for the mental model developed would also lead the designer to reconsider removing (automating) the last step of the process; as the anonymous feedback tells us the process is unintuitive.

Scenario 2

The second scenario was reported by one of the experimenters involved in the evaluation of the software. They joked that they had observed users sticking post it notes onto the screen with supplementary information. It is highly likely that a simple easy method to add additional textual information to the map was not explicitly considered in the development of the tool. Human beings are extremely efficient at finding 'work-arounds' when posed with problems. The method available in the digital system to add supplementary information was so convoluted that it was far simpler when faced, with time pressures, to stick physical post-it notes directly onto the screen. The metaphoric solution to this would be to create a digital post-it pad where information can be added and attached to geographical information; this would mean a central store of all data and allow the map to be paned without losing the post it notes geographical orientation.

Developing the Interface

Figure 3.18 is a vector graphics representation of the traditional bird table in Figure 3.17 the key elements have been recreated, the map; acetate pens; acetate overlays; post-it notes; book of 'stickies'; and communication methods. This graphical representation of the real world was then used to develop an interface (see Figure 3.19). The proposed solution is not intended as a final product rather it is presented as the first step in an iterative process for the development of a more intuitive interface display. The interface is designed around the principles of metaphoric references and direct manipulation. The interaction with the interface complies with standard practices. This simple proof of concept program was developed using 'Flash'. Objects are activated using a mouse over and then selected using a mouse click.

Figure 3.19 show the interface with the user using the pen. To change the colour the user simply selects the desired coloured pen. Perspective has been added to the objects to show when they are in use, or have been picked up. This use of perspective not only shows that the object has been selected, but it also allows users the ability to 'mouse over' symbols for a better look (see Figure 3.20).

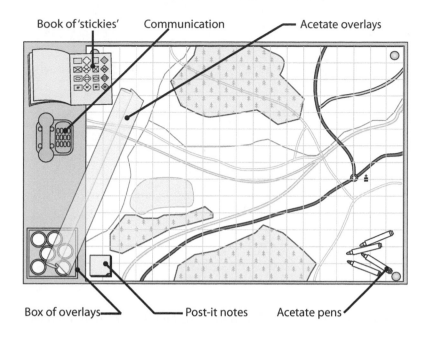

Figure 3.18   Graphical representation of bird table

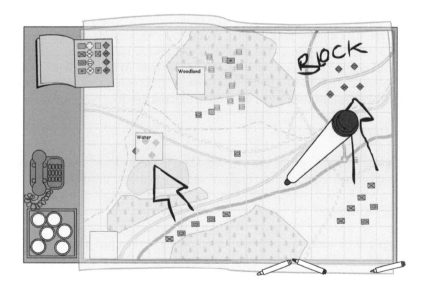

Figure 3.19   Screen shot from demonstrator tool showing pen in use

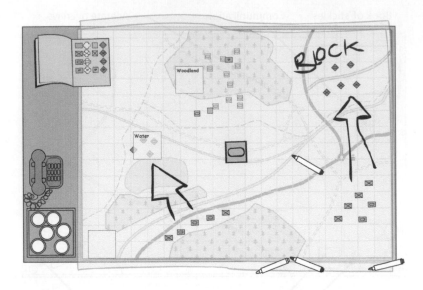

Figure 3.20   Screen shot from demonstrator tool showing unit highlighted

Adding Units

A more intuitive interface allows the user to drag the physical objects around the screen. The user simply positions the mouse over the desired object, depresses the left mouse button, drags the mouse to the desired location and releases the key (direct manipulation and standard practice for both Apple and Windows OS). As the objects are activated by a 'mouse over', they enlarge to give a sense of perspective and allow the user to view them in more detail

Finding the Correct 'Sticky'

To select the required 'sticky' the user would go to the book and select the correct page, the 'stickies' could then be dragged and dropped onto the map.

Drawing on the Map

Drawings can be added by simply clicking on the desired pen analogous to pick it up; this is used to draw by depressing the mouse. Information added would automatically be attached to the highest-level acetate.

Feedback

Whilst no formal testing of the interface was conducted, the interfaces intuitiveness was explored by asking users to perform simple tasks immediately after being presented with the display for the first time. The display closely mirrors the physical environment and follows standard approaches such as drag and drop mouse manipulation. As a result, users had no problem in completing these tasks such as adding units to specific geographical location, and adding an overlay.

Conclusions

Metaphors need to be applied to the interface in a considered manner; a poor metaphor can lead to confusion if an incorrect mental model is created by the user. Any interface needs to be tested with the end user and the information collected by observation of use. Mohnkern (1999), comments that while the metaphor is a useful way of providing consistency and structure, and of providing an initial set of functions and affordances to work with, we need to be bold when it comes to breaking from the metaphor when necessary (when the interface needs to perform in a manner inconsistent with the metaphor).

Chapter Summary

This chapter has introduced some of the basic principles behind the 'art' of interaction design and, more specifically, interface design. The chapter has shown that displays can be constructed based upon their individual information requirements; a set of design heuristics is presented that informs the representation choice of this information. Basic design principles for understanding and interaction are also discussed. These principles are then explored further in a military battlefield-planning case study. Norman (2007) offers six overriding design rules; these provide a concise summary for much of the work described in this chapter:

1. Provide rich, complex, and natural signals.
2. Be predictable.
3. Provide a good conceptual model.
4. Make the output understandable.
5. Provide continual awareness, without annoyance.
6. Exploit natural mappings to make interaction understandable and effective.

The chapter is, in no way intended to be exhaustive; its aim was to provide a basic grounding for the subsequent chapters where design is integrated with the CWA framework. As stated in the introduction to this chapter, a more systematic approach to design is required for complex sociotechnical systems. Whilst the design principles discussed in this chapter still apply; a more structured approach is required to capture the complex relationship between system constraints. In the remainder of this book CWA is presented as a process for modelling these constraints. CWA is based upon a theory of ecological psychology; according to Burns and Hajdukiewicz (2004) the three main tenets of ecological psychology (that influence design) are.

1. People's actions are constrained by their environment or work domain, so that the work must be understood before starting a design.
2. It is possible to design interfaces (or 'mediated environments') that provide information that people can pick up and use.
3. There are visual ways of displaying information that can reduce the need for memory or mental calculation.

It is therefore proposed that ecological approaches such as Work Domain Analysis are suitable for considering these tenets. Unlike many other tools, EID/WDA supports the requirement that interaction design should be technologically agnostic. Lintern et al (2004) offer a list of four, more specific, points to consider when designing an interface or workspace:

1. Information Requirements Definition – Determine 'what' information needs to be displayed.
2. Information Layout – Determine 'where and when' information should be presented, relative to other information in the display.
3. Visual Representation Design – Determine 'how' components of information can best be represented, so that operators can rapidly perceive meaning from the information display.
4. Navigation and Linking – Determine how a worker can navigate through the information space and how they can integrate pieces of information that need to be associated.

Lintern et al (2004) go on to classify the phases of CWA to how they meet the four requirements:

'The Work Domain Analysis (WDA) identifies essential functions and thereby the specifications for information that must be displayed to represent those functions. By showing how different functions need to be associated, this analysis also provides specifications for access, navigation and linking between items of information.

The Control Task Analysis (ConTA) maps relationships between control tasks that reveal sequential patterns of information access, thereby complementing the specifications for access, navigation and linking from the Work Domain Analysis.

The Strategies Analysis (StrA) maps alternative ways of performing control tasks which identifies different frames of reference, alternative process flows, and alternative patterns of access to information, thereby identifying further requirements for access, navigation and linking.

The Social-Organisational Analysis (SOCA) maps patterns of coordination and communication between workers and also between workers and automated agents, thereby identifying requirements for communication links and, by identifying the content and amount of information to be passed on that link, identifying some essential capabilities of that link.

The Worker Competencies Analysis (WCA) identifies the level of cognitive control (knowledge, rule or skill) at which a work element is typically performed, thereby suggesting the representational form for the relevant information.'

Chapter 4

Application of CWA in Familiar Domains

Chapter Introduction

As discussed in Chapter 2, CWA was developed to handle complex sociotechnical systems. In order to explain the analysis of these systems it is important the reader first has (or is given) an understanding of the environment; with such complex domains this is often no mean feat. The normal approach in this situation would be to use a simplification of the domain; however, in this case if the domain is simplified many of the reasons for having CWA in the first place become obsolete, despite this; it is perceived that a chapter describing a more simplistic model is of benefit for explaining the process.

An iPod

This case study uses the first three phases of CWA to explore the constraints governing the use of the Apple iPod. The analysis describes the constraints governing how activity is conducted in an actor independent way. This approach is used to highlight how system constraints limit functionality in specific situations. This understanding then forms a structured basis for designing out these limitations. To explain the approach a familiar, simplistic domain is used that of an 'Apple iPod'. This domain has been selected to communicate the approach, it is contended that whilst this domain lacks the complexity of larger sociotechnical systems traditionally analysed with CWA, it is complicated enough to serve as an example to introduce the approach.

Introduction

As stated in the introduction, the example used to describe this approach is that of an Apple iPod, the specific product has been selected due its popularity and predicted familiarity to most readers. The iPod is a portable digital music player (see Figure 4.1) that allows users to store up to 20,000 songs (dependant on model; figures based on the 80GB version) in a single portable product. The product, measuring 103 × 61 × 14 mm (see Figure 4.2), is not much larger than an average mobile phone, allowing it to fit comfortably into most pockets. Stored music can be played back via the supplied headphones, a computer, hi-fi or car stereo (with appropriate leads or third party gadgets, see Figure 4.3). The first iPods were initially developed solely as digital music players; video was later added as an additional feature. The following case study documents the analysis of a 4th Generation iPod (see Figure 4.1); this subsequently updated version of the iPod was the last to remain solely as music player before the introduction of the colour screen and video playback capabilities.

The success of the iPod is irrefutable; since its introduction in October 2001, it has continued to grow in popularity. Claiming a market share in excess of 62 per cent (Jobs, 2007a; based on data for November 2006), the iPod is, without doubt, the world's best selling digital music player. This success is likely to have been influenced by a number of factors including: Apple's strong brand; the product's ease of use; the product's strong design style; and a string of successful marketing

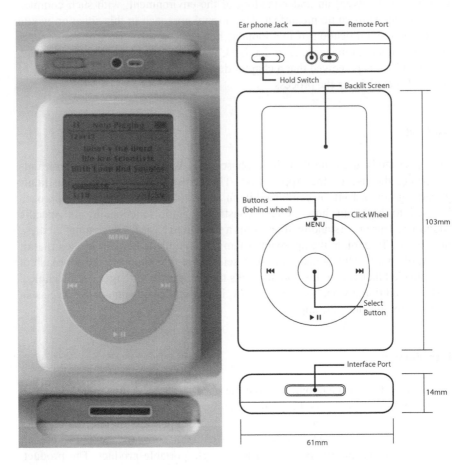

Figure 4.1   4th Generation iPod (Photographed by the author)

Figure 4.2   External attributes of an iPod

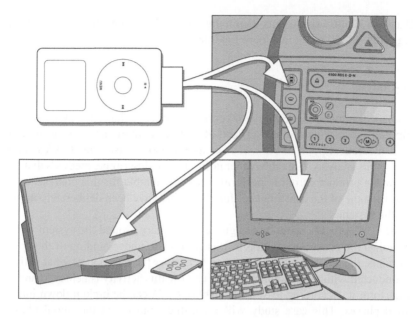

Figure 4.3   Common products the iPod is interfaced with

campaigns. Apple's design ethos appears to strike a good balance between a focus on the products technical capabilities and what Redström (2006) terms 'the user experience'; delivered through the software and products interface design. From a consumer perspective the iPod is clearly a great product; not only does it dominate the sales market, unsurprisingly, the iPod also receives consistently high reviews in independent opinion polls and magazines product comparison articles (T3, 2006; The Gadget Show, 2007). It is also assumed that the Apple organisation have also done exceptionally well from the product, even without considering the direct revenue from sales of over 100 million units (announced in April 2007; Apple, 2007a), Apple have massively raised their profile and brand awareness. It is postulated that Apple's decision to make iPods solely compatible with iTunes has also had significant implications for the success of the organisation; a huge number of PC's now have Apple software sitting on them, many, for the first time. The free to download software is responsible for over 2 billion music track downloads (as of January 2007) equating to sales of 5 million songs a day; when these downloads are considered along side CD sales, Apple are the 4th largest music reseller in the US (Jobs, 2007a). Initial speculation would suggest that the Apple organisation has done an outstanding job of striking a balance between creating a product that is a great design for the user as well as a great design for the growth and profitability of the organisation. With this in mind, such a successful product may not seem the obvious choice for a chapter on system improvement; however, as this chapter will show despite the success of this product, the systematic approach discussed is still able to identify constraints within the system that can be manipulated through design; that would result in a more flexible system.

The Analysis

The analysis aims to highlight the level of flexibility within the system (the iPod, computer, accessories and human) and identify any future areas for development from a theoretical point of view. Due to its focus on constraints, CWA is ideally suited to this analysis. Arguably, other normative or descriptive methods that focus on how the product currently performs are less suitable for this analysis as they may miss some of the constraints or functions that have been deliberately imposed upon the system.

The CWA process starts by assigning a boundary around the system in order to establish the constraints within it. Based upon an understanding of the domain constraints, the analyst is then prompted to answer the question of what activities are conducted within the domain as well as how this activity is achieved and who can perform it. According to Vicente (1999), CWA can be broken down into five defined phases. This case study will focus the analysis on the initial phase of CWA, Work Domain Analysis, the subsequent three phases; Control Task Analysis and Strategies Analysis will also be used. In this particular case, the benefits of extending the analysis into the last phase, Worker Competencies Analysis were not perceived to be great enough to warrant the analysis.

Work Domain Analysis

The way that WDA is conducted is dependent on the system; in the case of the iPod, we have a good understanding of the physical components and we can quite easily extract the functions that these components can afford. It is also possible to assume the overall purpose of the system. An Abstraction Hierarchy (see Figure 4.5) can be used to model these assumptions and to create a link between what the system is capable of, and what the system purpose is. At the top of the hierarchy the overall purpose of the system is recorded, at the base of the hierarchy the physical objects are recorded. Between these levels, the system is described at a number of levels of abstraction (typically five); these levels are linked using structural means-ends-links

The Physical Objects

The 'physical objects' are represented at the base of the hierarchy; at this level the components that make up the iPod both internal and external are recorded. These components can be seen graphically in Figure 4.2 and Figure 4.4.

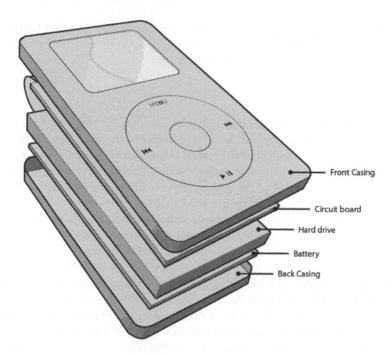

Front Casing

Circuit board

Hard drive

Battery

Back Casing

Figure 4.4   Exploded view of an iPod showing internal components

It is also worth considering the other objects that exist in the iPod's domain that make up significant parts of the system. In this case, a personal computer is an integral part of the system, without a computer, it is impossible to upload (add) music on to the iPod. Many users would also consider objects such as car stereos and hi-fi systems to be integral parts of the system.

The Object Related Processes

The 'object related processes' level resides above the 'physical objects' level, the object related processes capture the affordances of the physical objects, these processes should be independent of the purpose of the system and recorded in very generic terms. An example of a function for the screen would be 'to display textual and graphical information' not 'to show track number', this generic function description is important as it allows the screen to be considered for purposes other than the display of track information.

By adding links between the objects and the purposes they can afford it is apparent that some components have more than one affordance. It can also be observed that some functions require more than one physical object to perform. Some functions can be conducted by more than one object, thus, indicating some level or redundancy. As an example, the interface port can be seen to allow a number of interface related functions to be carried out. In addition, the function of converting electrical signals to sound can be conducted by a number of physical objects including headphones hi-fi and Car Stereo.

Functional Purpose

The 'functional purpose' level is the highest level within the Abstraction Hierarchy. The task of determining the functional purpose(s) of the iPod is a fundamental consideration, this will influence the rest of the hierarchy, the purpose set should be independent of any particular activity and it should capture the reason for the design or procurement of the system. By examining the functionality of the first generation iPod, it could be assumed that the product was developed for the sole purpose of consolidating a music collection into a portable device. This purpose focuses quite heavily on a physical user-centred definition of the iPod. There is no consideration to any of the purposes placed by the Apple organisation relating to brand image, revenue or market share that are important to almost all business organisations. Without modelled these constraints, this analysis would be incomplete; there appear to be constraints placed upon the iPod that cannot be explained purely by a user centred definition of the product. There seems to be no real reason from a technical perspective why the iPod should work solely with Apple's own software (iTunes); many of the iPod's competitors do not have this constraint, they allow their products to interface with a host of software packages. Another key constraint on the iPod is the ability to copy and transfer the music from one computer to another. From a technical perspective this should be possible as it is in many other digital storage devices; however, within the iPod this is a very difficult thing to do. According to Jobs (2007b), much of the complexity of this issue lies within the contracts between Apple and the 'big four' music companies: Universal, Sony BMG, Warner and EMI, since Apple does not own or control any music itself, it must license the rights to distribute music from others. These music companies as part of their agreement to allow music to be distributed electronically enforce a requirement that the downloaded music be protected limiting its distribution; in Apple's case to five computers. The agreements in the contracts between Apple and the music companies, explains the constraint that music downloaded through iTunes can only be played on an iPod and the constraint that legally protected downloads from other services are not compatible with the iPod. However, it does not fully explain Apple's decision to make the iPod solely compatible with iTunes, as Jobs (2007b) points out an estimated 97 per cent of the music on the iPod is unprotected.

It seems to be apparent that there are additional factors influencing the decision to add in these constraints to the iPod system, other than a user centred design. From a user perspective the purpose of the iPod could be captured as 'to provide mobile acoustic entertainment' as well as 'to consolidate and develop music collection'. From an Apple organisation perspective the overall purpose could be assumed to be 'to increase growth and profitability for the organisation', this could be broken down to 'increase iPod sales' and 'increase iTunes song sales'.

Values and Priority Measures

The values and priority measures level sits directly below the overall functional purpose level, here information is captured that can indicate how well the system is achieving its functional purposes. The criteria here should be defined in terms that can be measured, for example; the capacity in Gigabytes or subjective and performance ratings for 'ease of use'. The values and priority measures are linked to their relating functional purposes using means-ends links. Portability is an important function in providing mobile acoustic entertainment; this in turn increases iPod sales. The Ease of use of the product influences all of the functional purposes, with the exception of 'increase iTunes song sales'. The ease of music sharing is important in the development of music collections. Capacity has an influence on all of the function purposes, a suitable capacity is required for acoustic entertainment, for developing and consolidating a music collection, increasing iTunes sales and increasing iPod sales. An insufficient capacity would have a detrimental effect on each of these functional purposes. Finally, the user experience effects the acoustic entertainment as well as iPod sales.

Purpose Related Function

The purpose related function level sits in the middle of the hierarchy; this level links the purpose independent functions at the base of the hierarchy to the more abstract ideals of the system at the top of the hierarchy. Using means-ends links the AH can be connected up; any cell can be selected as what is trying to be achieved, the connected cells above should answer the question of why this should be achieved, the connected cells below should answer the question how it is achieved.

From looking at the hierarchy, a partial conflict can be observed between the user's functional purpose of developing their music collection and the Apple organisation's perceived requirement to increase music sales through iTunes downloads. Technologically there are a number of additional ways in which a user could easily increase their music collection, this could be done by making it easier to copy music across from a number of computers; another method would include some level of synchronisation between iPods either by cable or wireless technology. It is postulated that reasons for Apple not including this level of functionality are also due to music piracy laws; however, other data storage devices and digital music players do not have these restrictions in place. It is also hypothesised that the organisations profit as a result of iTunes downloads was a significant factor in this decision.

Control Task Analysis

We have seen the benefits of viewing the domain independent of activity in the Work Domain Analysis; captured in the Abstraction Hierarchy. In order to understand the domain further it is advantageous to look at the known recurring activities. The second phase of the analysis, Control Task Analysis (ConTA), models these known recurring tasks focusing on what has to be achieved independent of how the task is conducted or who undertakes it.

The representation selected for this phase is the Contextual Activity Template (see Figure 4.6). The functions for the CAT, shown down the left-hand side, have been taken from the purpose-related functions level of the Abstraction Hierarchy (see middle row of Figure 4.5). In this case, the situations, shown along the top of the CAT, have been defined by how the iPod system is configured at any one time; the particular situations will have an effect on the functions that can be conducted. The situations under consideration are; the iPod connected to the mains on AC charge; the iPod using headphones whilst on the move; the iPod synchronised with the computer running iTunes; the iPod synchronised with a hi-fi system; and the iPod synchronised with some form of In Car Entertainment (ICE).

The dotted boxes in the template indicate where activity can take place, whereas, the whiskers indicate where activity typically takes place. Taking the top function 'provide power' as an example it can be seen that the iPod can receive

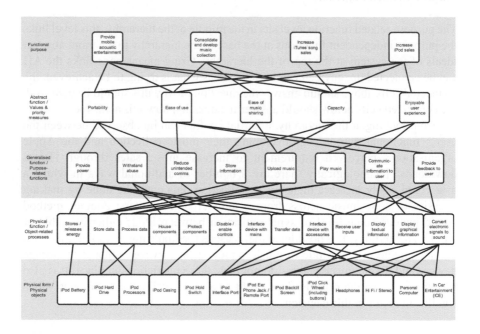

Figure 4.5   The Abstraction Hierarchy for the iPod system

power to charge the battery in all situations except on the move. The situations 'synchronised with hi-fi' and 'synchronised with ICE' are shown with only a dotted line to indicate that they can receive power in these situations; however, unlike whilst connected to a computer or connected to the mains, they may not typically receive power dependant on how the system is configured. Further development of the system into how the battery could be charged whilst on the move may be of significant benefit to many users of this product who have limited access to the situations where the iPod battery can be charged (i.e. mains, computer, hi-fi or ICE).

From Figure 4.6 it is also apparent that the only situation where music can be uploaded to the iPod is whilst connected to a computer. We can gain further information about why this is by revisiting the constraints captured in the Abstraction Hierarchy (Figure 4.5). By examining the links down from the node in

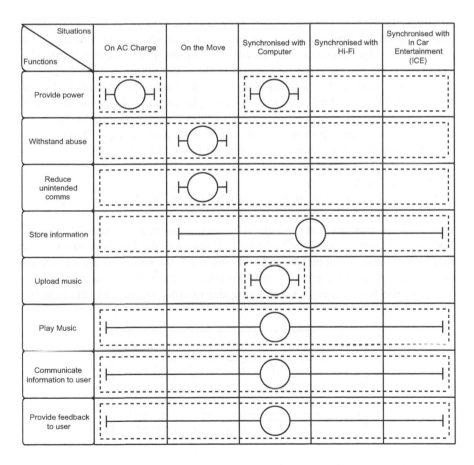

Figure 4.6   Contextual Activity Template for the iPod system

the centre of the Abstraction Hierarchy 'upload music' it can be observed that the only way of achieving this is to use a combination of the personal computer and the interface port. This current set up limits the iPod suitability to people who own or have regular access to a PC that they can load music on to. If the iPod were to contain functionality to record directly from a hi-fi or a Car stereo then the product suitability could be extended.

The remaining functions show a high level of flexibility within the system across the defined situations, many of the functions (play music; communicate information to user; and provide feedback to user) can and typically occur in all situations. The iPod can 'withstand abuse' and 'reduce unintended comms' at all times although typically this is only required when the user is 'on the move'.

Strategies Analysis

Strategies Analysis is used to look in more detail at known recurring activities. This stage of the analysis considers the tasks analysed in the ConTA phase and considers the strategies that are likely to be used to complete them. The definition of the activities can be taken from the CAT; although, sometimes, it is advantageous to describe these activities in terms that are more specific. Figure 4.7 shows a Strategies Analysis for the activity 'play music' (taken from the CAT), here the tasks are described in more detail to show the steps required to get from the start-state 'desire to hear music' (shown on the left of Figure 4.7) to the end-state 'listening to music' (shown on the right of Figure 4.7). The analysis does not aim to prioritise or comment on the best strategy, it merely aims to capture the flexibility within the system.

Figure 4.7 shows the Strategies Analysis used to describe the possible situations defined by the constraints in the system; it is also possible to use the same representation to start thinking about new strategies for extending system flexibility. Returning to the CAT, shown in Figure 4.6, the functions that require additional flexibility can be identified. In Figure 4.8, the strategies are considered for the task of 'provide power'; more specifically the task has been refined to 'charging the iPod battery'. Considering the existing constraints, the possible

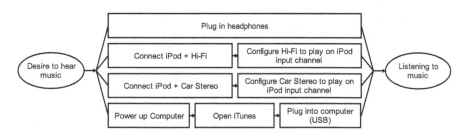

Figure 4.7   Strategies Analysis for play music

strategies are shown at the top of the diagram in white boxes. In the shaded boxes, included at the base of the diagram, new strategies for developing methods of charging the iPod batteries whilst on the move are included. These strategies rely on the introduction on new physical objects, or technologies, into the system; in this case third party accessories that could include solar panels or some kind of motion-to-electricity transformer possibly fitted to footwear.

Figure 4.9 shows the Strategies Analysis for uploading music, the white boxes show two of the current strategies, bound by existing constraints, for uploading music. Both methods require the user to connect the iPod to a computer with iTunes installed on it. The first strategy shows a system configured to synchronise the iPod with the computer automatically; the second strategy is for a manual update. Three additional strategies are shown as shaded boxes; these strategies require manipulation of the existing constraints, in this case a change to the iPod's software. The strategies cover recording audio files directly from a hi-fi or car stereo, this additional functionality would allow users to plug their iPods directly into a hi-fi or car stereo output and record tracks from other formats, such as vinyl records, directly on to the iPod. A further strategy is also included at the bottom of Figure 4.9, this strategy allows two users to connect their iPods together. Once connected the iPod could show non-duplicate songs, the users could then select songs from this list that they wished to copy across. The connection could be made physically by a cable or wirelessly via a Bluetooth or WLAN technology.

Figure 4.10 shows a CAT for an iPod system with the additional functionality described in Figure 4.7 and Figure 4.8 included, unlike the CAT for the original system (Figure 4.6) the proposed system allows all functions to be completed in all situations, thus extending the flexibility of the system. The mini-strategy-flowcharts in Figure 4.6 show the strategy required to extend the flexibility of the system. The task of 'provide power – on the move' can be completed using either of the bottom two strategies in Figure 4.8. The tasks of 'upload music – on AC charge' and 'upload music – on the move' can be completed using the bottom strategy in Figure 4.9. The task of 'upload music – synchronised with hi-fi' can

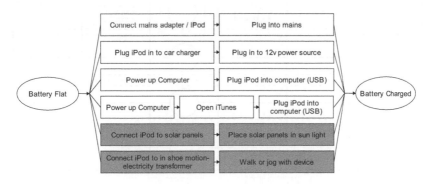

Figure 4.8   Strategies Analysis for charging the battery

be completed using the middle strategy in Figure 4.9. Finally, the task of 'upload music – synchronised with ICE' can be completed using the fifth strategy down in Figure 4.9.

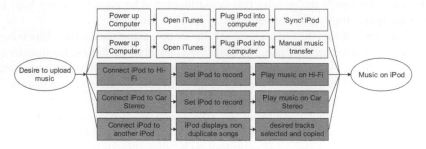

Figure 4.9   Strategies Analysis for uploading music

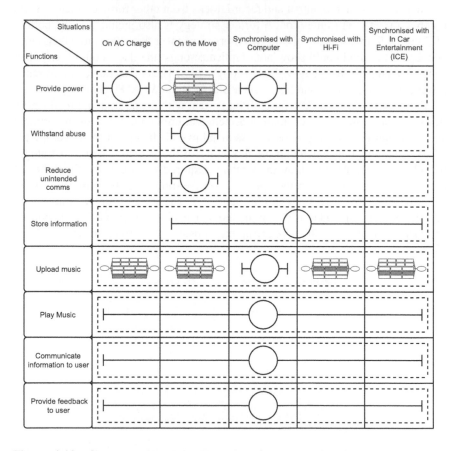

Figure 4.10   Contextual Activity Template for the modified iPod system

Freedom and flexibility within a system allows users to adapt and select a way of achieving an end-state that is most appropriate to them in a given situation. A flexible system is also likely to support and promote formative thinking and emergent behaviour. In some cases, a tightly constrained system, with only one way of achieving an end-state, may be beneficial; the presence of a choice requires decision-making, this can slow down the process of conducting the task. A well-developed interface design, which presents the most common strategy clearly, whilst still allows expert users to adapt their behaviour is often the best way to tackle this conflict.

Conclusions

In this case study the initial three phases of CWA are used to represent the constraints affecting what functions are possible within the system, as well as which functions can take place in which situations, and how. A clear link is shown between the 'purpose related functions' in the Abstraction Hierarchy and the 'functions' in the Contextual Activity Template. This link allows analysts to use the means-ends relationships in the AH to explore the reason behind some of the constraints in the CAT. The 'functions', from the CAT, that are possible with the existing constraints in place can be explored further in the Strategies Analysis phase of the framework. This approach then uses the same representation to explore ways of manipulating these existing constraints to allow functions to be conducted in new situations. A number of system design changes were proposed that would extend the flexibility of the system. The proposed changes would allow all of the functions to be completed in all of the modelled situations.

This analysis has modelled the iPod system in terms of the constraints governing its use and operation. In the WDA the approach has described the relationship between what the physical objects in the system can perform, and what the system is required to perform. In the ConTA, the approach goes on to describe situations in which the iPod is constrained in its functionality, such as charging the battery on the move and uploading music whilst away from a computer. Looking at these functions, or activities, in more detail it has been possible to suggest potential strategies for removing these constraints. From interpretation of the captured constraints, it is possible to postulate some of the possible reasons for their existence in the analysed system. As stated in the introduction to this chapter; it is evident from the sales figures of iPods, independent product ratings and the number of track downloads that Apple seems to have struck a successful balance between the requirements of the user and the requirements of the organisation. Putting aside the requirements of the Apple Organisation for one moment, the analysis makes it possible to suggest some changes to the design of the iPod system that enhance flexibility; these changes would remove some of the constraints placed upon the operation of the iPod, providing its users with additional flexibility and functionality. It is predicted that some of these changes, specifically those related to music sharing, would have an effect on the hypothesised requirement of the Apple organisation 'to increase sales of iTunes downloads'. The increased functionality to support the sharing of music would have a significant impact on allowing users to develop their music collections. As Figure 4.9 shows, of particular benefit would be a system that could determine and display non-duplicate songs possibly through an iPod-to-iPod music transfer function, such as that built into the 'Microsoft Zune' music player. Landmark cases such as the agreement between Apple and EMI (Apple, 2007b) to remove DRM (digital rights management; the protection encoding that stops users placing the music on more than five computers) indicate that the music industry is rethinking the strict restrictions it places on Apple and its competitors. Only time will tell if Apple choose to open up the iPod to other music management software, exposing if the decision to limit the iPod to iTunes was solely due to copyright agreements with the music companies or an attempt to entice more customers into the iTunes music store.

In summary, this chapter has introduced a new, structured, approach for exploring the constraints limiting system functionality in different situations. Further, the approach has highlighted the relationships between the initial three phases of CWA and how these can be exploited to aid the manipulation of system constraint to extend functionality and, in turn, flexibility. A simplistic domain has been used, in order to communicate the approach without having to explain, in detail, the complex constraints commonly found in large sociotechnical systems. Whilst the described approach represents a new application of the existing CWA framework, it does not defer from the accepted definition. For this reason, it is contended that this approach is entirely scalable to the larger sociotechnical domains traditionally analysed with CWA

Chapter Summary

This chapter has discussed the application of CWA in a simplistic, familiar domain, as outlined in the introduction to this chapter; it is acknowledged that the framework's applicability in simple domains can easily be questioned. As the complexity of the domains increases, the models grow in size and the benefits of CWA start to become apparent.

The relationship between the initial three phases of CWA is explored. A clear link is shown between the 'purpose related functions' in the Abstraction Hierarchy and the 'functions' in the Contextual Activity Template. This link allows analysts to use the means-ends links in the AH to explore the reason behind some of the constraints in the CAT. The 'functions' in the CAT can be explored further in the Strategies Analysis phase of the framework. The StrA phase was used to consider strategies for extending the flexibility within the CAT. A number of system design changes were proposed that would extend the flexibility of the system. The proposed changes would allow all of the functions to be completed in all of the modelled situations.

Chapter 5

Applications of CWA in a Complex World

Chapter Introduction

By extending the analysis into more complex environments, this chapter aims to build upon the theoretical explanation of CWA, presented in Chapter 2, and illustrated in the familiar example in Chapter 3. The previous chapter focused, primarily, on an artefact based level; this chapter aims to address the ability of CWA to model 'systems of artefacts'. The first example focuses on the analysis of a system; the example is presented with the 'Rotary Wing Military Mission Planning System' this case study investigates the application of the first four tools of CWA. The analysis of the 'Rotary Wing Military Mission Planning System' starts to move beyond a description of the systems, broad design recommendations are drawn from the analysis. The second, case study in this chapter looks at how CWA can be used to evaluate these complex systems. The example of a battlefield command and control system is used. A technologically agnostic model is created; this then forms the basis for evaluating the effects of new technology (digitisation) on different levels of system performance.

Design – CWA of Rotary Wing Military Mission Planning System

Introduction

The aim of this chapter is to illustrate how a more complete CWA can be used to extract broad design recommendations for system improvement. The case study used to highlight this is a CWA of a generic Mission Planning System (MPS) software tool. The purpose of the examined MPS is defined as 'to support mission planning for various military rotary wing aircraft including; attack, transporter, reconnaissance, and surveillance helicopters.' The analysis starts by developing a constraint based description of the system; from this description a number of broad design recommendations are extracted; these relate to the future development of the MPS software tool, with particular attention to interaction design through the organisation of on-screen data. The process of conducting the analysis also elicited a number of additional benefits; these include the information required for the redevelopment of the training syllabus structure, informing lesson sequencing and teaching.

Mission planning is an essential part of flying a military aircraft. Whilst in the air, pilots are required to process, in parallel, cognitively intense activities including; time keeping, hazard perception, and off-board communication. These activities are all conducted whilst attending to the task of navigating through a three-dimensional airspace. Pilots are required to evaluate the effects their actions have on others within the domain constantly. Decisions need to be made that consider any number of both military and non-military services, organisations and civilian groups. Calculations need to be made based upon a number of physical considerations, these include; environmental constraints, aircraft performance and payloads. Pilots also need to balance mission objectives with rules of engagement and high order strategic objectives. Pre-flight planning is one essential method used to alleviate some of the pilot's airborne workload. This planning process, which was formerly conducted on paper maps, is now supported by a digital software based planning tool. The MPS software tool described is currently used by the UK army to develop and assess mission plans for attack helicopters. The MPS software tools provides and processes digital information on; battlefield data, threat assessment, intervisibility, engagement zones, communication details, transponder information, and IFF (Identification Friend or Foe) settings. In short, the MPS is used to plan and assess single and multiple aircraft sortie missions. Whilst for the purposes of this chapter, a specific MPS tool was used, it is contended that the analysis could apply to many other software based mission-planning tools in both military and civilian domains.

Mission plans are generated prior to take off on PC based MPS terminals. Key information developed in the software tool is transferred to the aircraft via a digital storage device called a 'Data Transfer Cartridge' (DTC). Information is presented on the Aircraft's onboard flight display. This multi-function display can be used by the pilot for to assist in navigation and target identification. This process is represented graphically in Figure 5.1.

The digitisation of the planning process has a number of benefits; by performing multiple parallel calculations, the computer is able to consider a huge number of variables that would be inconceivable in a paper-based system. When combined with complex algorithms, this allows for greater accuracy in modelling factors such as fuel burn rates. The design of the user interface for the software system has the potential to affect the performance of the operators significantly. The visualisation of the plan is constrained to a limited screen real estate. Therefore, the navigation

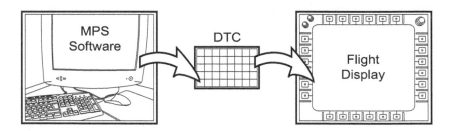

Figure 5.1   The planning process transferring information between terminal and the aircraft

and clustering of data need to be carefully considered. The design of these digital systems needs to be contemplated in light of new constraints and freedoms.

Based upon the new capabilities and constraints within a digital system it is possible to rethink task distribution. Activity can be distributed amongst the team through a simple network allowing tasks to be completed collaboratively. A number of approaches have been successfully applied in the past to model these interactions within command and control domains. These include Social Network Analysis (Houghton et al, 2006); Event Analysis of Systematic Teamwork (EAST; Walker et al, 2006a); and models of team situation awareness (Stanton et al, 2006; Gorman et al, 2006). These approaches tend to focus on current activity. The approach presented in this chapter aims to inform the design of future generations of the mission planning system though the use of an event independent analysis technique.

Why Cognitive Work Analysis?

The MPS system is used to develop plans in an extremely complex environment. We can gain some perspective of this, by considering it against Woods's (1988) four dimensions for complexity:

- Dynamism of the system: The system is extremely dynamic; it changes frequently without intervention from the user. Whilst control orders that govern the airspace are used to limit this dynamism, mission start times are often subject to change, thus making previous assumptions invalid.
- Parts, variables and their interconnections: There are a number of services and organisations operating within the airspace and ground environment. These groups often have competing aims and objectives.
- Uncertainty: Because of the 'Fog of War', data can frequently be erroneous, incomplete or ambiguous. This makes it difficult to make predictions about future events.
- Risk: Potentially, decisions made within the environment made have life and death consequences.

Based upon Woods's (1988) heuristics, there is no doubt that the environment the MPS serves is extremely complex. Zsambok and Klein (1997) describe battlefields as environments that have high stakes; are dynamic, ambiguous, time stressed and in which goals are ill defined or competing.

An approach is required to model the MPS domain that is independent of time or specific context. Normative analysis techniques focus on how the system currently performs, or how the system should perform. The models they produce are therefore, only applicable for specific examples, Jenkins et al (2008b) point out that these models soon become invalid as system parameters change. According to Naikar and Lintern (2002) normative approaches specifying temporally ordered actions, result in workers being ill prepared to cope with unanticipated events. For this analysis a formative approach was required that, through its focus on constraints would allows the analyst to exhaustively, but concisely, describe the system under analysis. Vicente's (1999) description of Cognitive Work Analysis (CWA) addresses these requirements. Although initially developed for closed-loop, intentional, process control domains; CWA has been successfully applied to a number of open-loop military systems (e.g. Burns et al, 2000; Naikar et al, 2003). Burns et al (2000) apply Ecological Interface Design (an approach evolved from CWA) to model shipboard command and control. They use this example to explore how the Work Domain Analysis (WDA) model can be extended to apply to open-loop systems with boundaries that are much harder to define than their closed-loop counterparts are. Burns et al (2000) justify the use of their approach by drawing upon similarities between decision making in naval command and control and the process control domains described by Rasmussen et al (1994) and Vicente (1999). Burns et al (2000) point out the safety critical nature of both domains as well as the underlying physical constraints.

The described analysis builds upon the work of Burns et al (2000) in exploring the appropriateness of CWA in open-loop complex systems. Burns et al (2000) limited their analysis to the initial phase of CWA (WDA). Whilst subsequent work in the same domain by Lamoureux et al (2006) as well as other command and control examples (Naikar 2006) extended this analysis to the second phase. There has been little attempt in the literature to extend the CWA framework beyond these two phases. The social and organisational analysis phase builds upon the products of previous phases. This analysis described involved constructing: an Abstraction Hierarchy, Abstraction Decomposition Space, and Contextual Activity Template for use within the SOCA phase. According to Rehak et al (2006) it is through a process of viewing the same domain in a variety of ways that many design innovations arise.

Data Collection

Access was granted to a number of Subject Matter Experts (SMEs) in order to provide the analysts with a high level of domain understanding within a limited period. The SMEs were made up of a combination of flight instructors and serving airmen. An initial two-day meeting was held in order to introduce the analysts involved to the helicopter mission planning process and the MPS software tool. The data collection process involved a number of interviews held with four different SMEs and SME walkthroughs of MPS planning tasks. In total, three interview/walkthrough meetings were held at Brunel University, each lasting approximately five hours in duration. Subsequent visits were also made to the army flying school based at Middle Wallop. The data collected during these sessions was used to create; the Abstraction Hierarchy (AH), Contextual Activity Template (CAT), Strategy Analysis flow chart, and Social Organisation and Co-operation Analyses (SOCA). Each analysis draft was validated by the SMEs and updated based upon the feedback.

Analysis Results

The social organisation and cooperation analysis (SOCA) phase builds upon the previous phases of CWA. The first three phases of the analysis are actor independent. The SOCA phase revisits the products produced, considering the constraints governing which actors can be involved with each activity. It is therefore important to consider the initial phases of CWA before considering the SOCA phase.

Work Domain Analysis

Work Domain Analysis (WDA) is used to describe the domain in which the activity takes place independent of any goals or activities. The first stage of this process is to construct an Abstraction Hierarchy (AH) of the domain in question. As previously stated, the AH represents the system domain at a number of levels; at the highest level the AH captures the system's raison d'être; at the lowest level the AH captures the physical objects within the system. The AH for the MPS is presented in Figure 5.2; in this case the system's functional purpose is 'To plan missions to enact higher command intent'; this is the sole purpose of the system in the current analysis. The second level down, values and priority measures capture the metrics that can be used to establish how well the system is performing in relation to its functional purpose; these include: Mission Completion (Adherence to Commander's Intent); Adherence to Rules of Engagement; Self Preservation; Minimise Unnecessary Casualties; Flexibility (adaptability); and the suitability of outputted data (DTC/UDM). At the very bottom level the physical objects that make up the system are recorded; in this case they are limited to the process of planning rather than the flight of the aircraft or the engagement of targets; examples include: maps and satellite imagery; orders; weather forecasts; flying regulations along with information on weapons, airframes and sights and sensors. The level above this object related processes captures all of the affordances of the physical objects for example from the Airspace Control Order (ACO) the airspace freedom and constraint can be elicited, as can the understanding of friendly unit's disposition and activity; at this level the affordances should be independent of the system purpose. The AH is linked together by the purpose related functions level in the middle of the hierarchy; this level puts the identified object related process into context with the system purpose.

Each of the levels can be linked by means-ends relationships; using the why-what-how triad. Any node in the AH can be taken to answer that question of 'what' it does. The node is then linked to all of the nodes in the level directly above to answer the question 'why' it is needed. It is then linked to all of the nodes in the level directly below that answer the question 'how' this can be achieved. Taking the example of payload required (see Figure 5.3), can first address the issue of why do we need to determine the payload required, by following the means-ends links out of the top of the node we can see that payload required is important

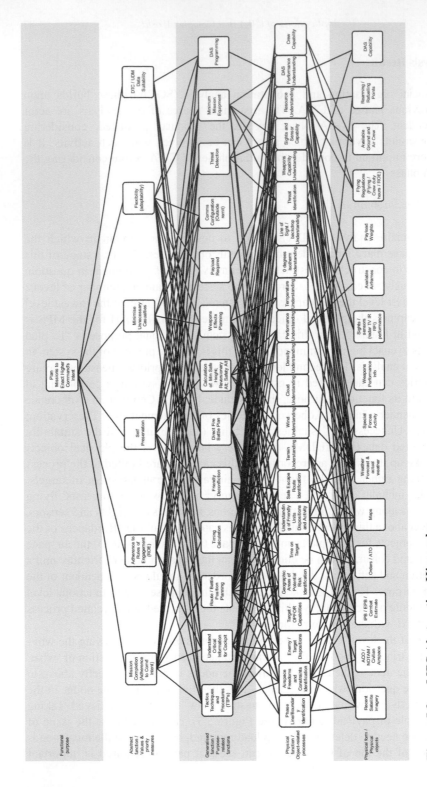

Figure 5.2 MPS Abstraction Hierarchy

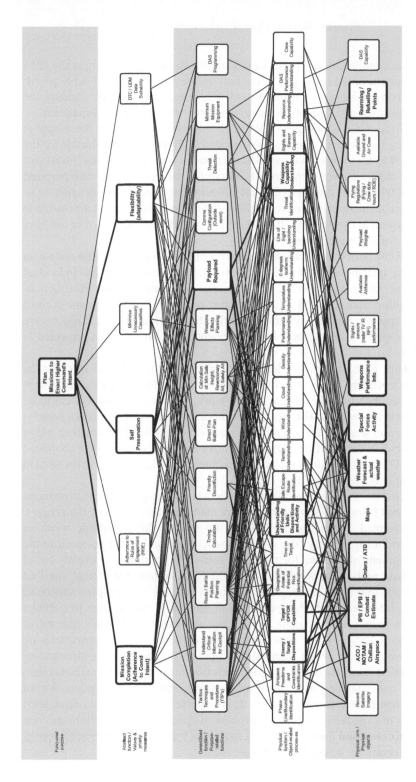

Figure 5.3 MPS Abstraction Hierarchy showing example of payload required

for; mission completion, self preservation and for flexibility. Looking at the links from the bottom of the node we can see how we determine the payload required; through having a weapons capability understanding, through understanding the enemy disposition and through understanding other friendly unit's disposition.

One of the main advantages of Work Domain Analysis is that the output is truly activity independent. The model generated in Figure 5.2 is applicable for the MPS as well as for the previous paper-based system. The objects in the lowest two levels may change as new technology is introduced; however, the system purpose, the way in which this measured and the object related processes are unlikely to change. By considering the hierarchy from a top down perspective, it is possible to view the system in a technologically agnostic way allowing the analyst or designer to conceive of a completely new system.

The AH can be decomposed based on levels of resolution through the system, in this case the system was decomposed into; total system, subsystem and individual components. Once decomposed, the data can be plotted on the Abstraction-Decomposition Space (ADS). The functional purpose(s) of the system in most cases will apply to the total system; similarly, the individual physical objects are likely to be either components or subcomponents. The MPS ADS is presented in Figure 5.4

The Work Domain Analysis leads the analyst and the SMEs to consider the domain independent of any activity taking place. This focus on why the system exists, rather than how the system should work, often enables the system to be considered in a new light. This new understanding can lead to significant benefits in the design of a new MPS system. It is postulated that the data structure and in turn the window design of the current MPS has been based on the data contained within the DTC rather than by the activity required; this has resulted in the need to have multiple windows open (see Figure 5.5).

By tracing the means-ends links within the AH, the design team can investigate task flow and information grouping requirements for each stage of the process. An AH informed design could eliminate the need to have multiple windows open to conduct an activity. The AH representation also has the potential to aid the development of training for the MPS software. Training is currently based on explaining each of the windows within the MPS system; the current interface design means that users are taught at an object level. A training plan structured into topics informed by the 'purpose related functions' allows users to consider why the functions are important, by tracing the means-ends links alternative approaches of achieving the same ends can be explored. This proposed training approach would result in new trainees developing a functional (i.e. understanding of the different functions involved and the relationships between them) rather than a physical understanding of the mission planning process (i.e. understanding of how each component window works). It is expected that this would have great advantages in expediting the training process by focusing initially on the mission planning process, and then on the MPS functions that support the process, rather than focus primarily on the MPS software tool.

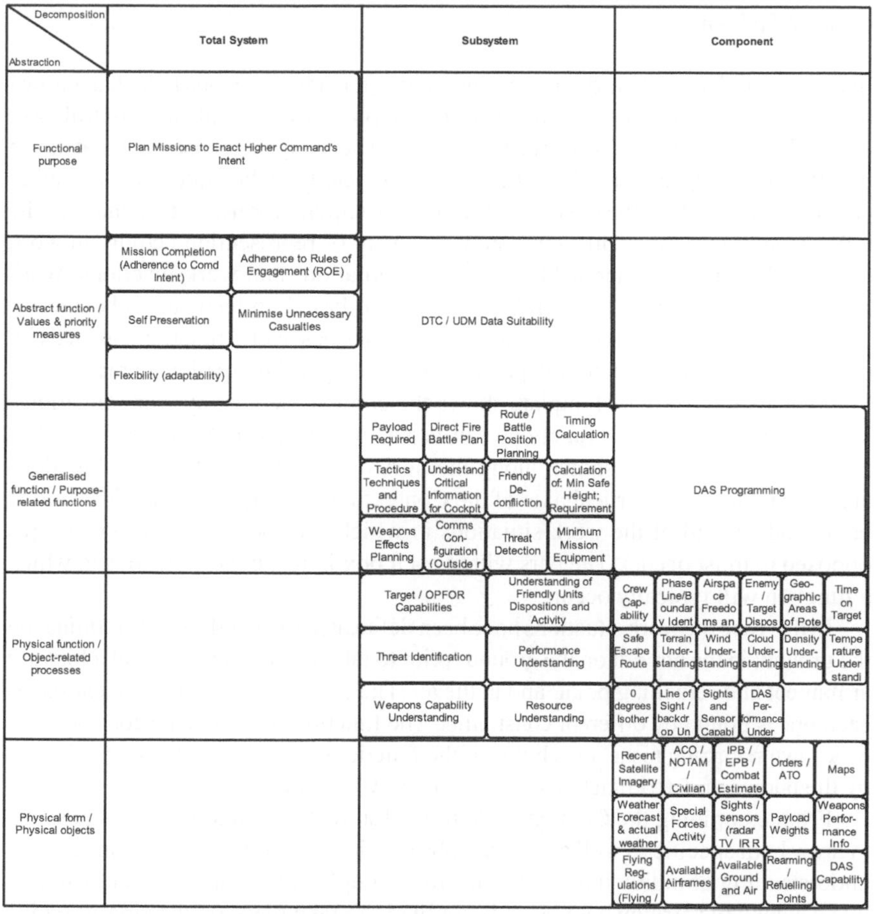

Abstraction \ Decomposition	Total System	Subsystem	Component
Functional purpose	Plan Missions to Enact Higher Command's Intent		
Abstract function / Values & priority measures	Mission Completion (Adherence to Comd Intent); Adherence to Rules of Engagement (ROE); Self Preservation; Minimise Unnecessary Casualties; Flexibility (adaptability)	DTC / UDM Data Suitability	
Generalised function / Purpose-related functions		Payload Required; Direct Fire Battle Plan; Route / Battle Position Planning; Timing Calculation; Tactics Techniques and Procedure; Understand Critical Information for Cockpit; Friendly De-confliction; Calculation of: Min Safe Height; Requirement; Weapons Effects Planning; Comms Con-figuration (Outside r; Threat Detection; Minimum Mission Equipment	DAS Programming
Physical function / Object-related processes		Target / OPFOR Capabilities; Understanding of Friendly Units Dispositions and Activity; Threat Identification; Performance Understanding; Weapons Capability Understanding; Resource Understanding	Crew Capability; Phase Line/Boundary Ident; Airspace Freedoms an; Enemy Target Dispos; Geographic Areas of Pote; Time on Target; Safe Escape Route; Terrain Understanding; Wind Understanding; Cloud Understanding; Density Understanding; Temperature Understandi; 0 degrees Isother; Line of Sight / backdr op Un; Sights and Sensor Capabi; DAS Performance Under
Physical form / Physical objects			Recent Satellite Imagery; ACO / NOTAM / Civilian; IPB / EPB / Combat Estimate; Orders / ATO; Maps; Weather Forecast & actual weather; Special Forces Activity; Sights / sensors (radar TV IR R; Payload Weights; Weapons Performance Info; Flying Regulations (Flying /; Available Airframes; Available Ground and Air; Rearming / Refuelling Points; DAS Capability

Figure 5.4   MPS abstraction decomposition space

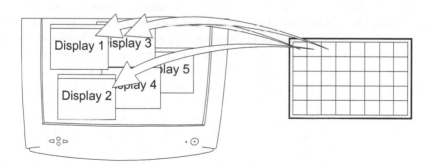

Display 1 isplay 3
lay 5
Display 2 play 4

Figure 5.5   Current structuring of MPS software is based on the DTC make up

Control Task Analysis

In order to understand the domain further, it is often advantageous to look at known recurring activities in more detail. The second phase of the analysis, Control Task Analysis, models these known recurring tasks. The analysis focuses on what has to be achieved independent of how the task is conducted or who undertakes it. Naikar et al (2005) introduce the Contextual Activity Template for use in this phase of the CWA (see Figure 5.6). This template is one way of representing activity in work systems that are characterised by both work situations and work functions. Work situations are situations that can be decomposed based on recurring schedules or specific locations. Rasmussen et al (1994) describes work functions as activity characterised by its content independent of its temporal or spatial characteristics. The Contextual Activity Template shows the context, defined by work situations, in which particular work functions can occur. According to Naikar et al (2005), the work situations are shown along the horizontal axis and the work functions are shown along the vertical axis of the Contextual Activity Template. The dotted boxes indicate all of the work situations in which a work function can occur (as opposed to must occur). The bars within each box indicate the situations in which a function will typically occur.

In this case the work situations have been delineated to include a MPS terminal on the ground; in the aircraft on the ground (prior to takeoff); on the ground at a forward armament and refuelling point; and in the air. The specific situations were chosen as each one is bound by different constraints. The functions captured are considered to be known recurring tasks, the choice of the functions can often be heavily informed by the purpose related functions level from the Work Domain Analysis.

From examining the Contextual Activity Template in Figure 5.6, it is possible to draw both specific as well as broader observations. Specific observations give an understanding of individual constraints, for example: target engagement planning, this can take place anywhere but it is not likely to take place on the aircraft whilst on the ground. It is also possible to build a broader image of the system by looking at patterns within the Contextual Activity Template: for example, it is rather salient that the only function that typically occurs in all situations is 'timing calculations'; this is due to the complexity of the system and the need for adaptation. It is also salient that in this domain all of the function can and typically do take place on the ground. Due to a number of mainly technical constraints some of the functions can only take place on the ground with the MPS system (such as calculations of safe heights; inter-visibly calculations; radar programming; resource allocation; understanding of critical information for cockpit; and determining the minimum mission equipment). However, other functions can take place in other locations but typically do not. From discussions with the SMEs it was clear the emphasis of planning is to get most of the functions completed on the ground leaving only minor alterations to take place in later situations. It can be seen from the dotted boxes that the majority of the functions (11 of the 17) can be conducted in all situations. Even with the additional capability and flexibility provided by a network-enabled

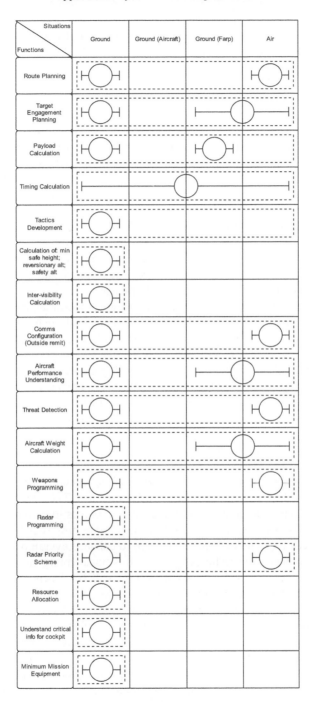

Figure 5.6   Contextual Activity Template

system, there still appears to be a strong emphasis on upfront, rather than on the fly, planning. By looking at the roles of actors in the SOCA phase, we will try to understand this further.

Strategies Analysis

Strategies analysis is used to look in more detail at known recurring activities. The strategy adopted by an actor at a particular time may vary significantly; different agents may perform tasks in different ways, the same agent might also perform the same task in a variety of different ways. There are often a number of different ways of achieving the same end state; often, the strategies use different resources and distribute the workload in different ways. This stage of the analysis has not been developed in any detail; however, an example has been included for completeness. The given example investigates ways of moving from an initial state, where the aircraft performance required is not achievable, to a final state, where the required performance is achievable. The strategies are provided in a random order, they are all strategies for increasing the aircraft performance.

The benefit of this representation is that it captures the strategies that can have an effect on a particular end state, the list may not always be exhaustive; however, it is intended to capture the main strategies that will have a significant impact on the outcome. It is possible that more than one of these strategies can be applied in parallel; the advantage of the MPS system over its paper-based predecessor is that the system is able to calculate the resultant effects of parameter manipulations on the rest of the system. In the paper-based system, these, often iterative, changes would take far too long to be conducted in a real life situation.

Social Organisation and Cooperation Analysis

Social Organisation and Cooperation Analysis (SOCA) addresses the constraints governing how the team communicates and cooperates. The analysis allows the constraints affecting the allocation of available resources to be modelled. In the vast majority of systems, it is desirable to determine how social and technical components can be combined and configured to enhance overall performance. In the case of complex socio-technical systems, this 'ideal configuration' is unlikely to be fixed; rather, the optimum configuration will be dependant on both the work functions and the work situation. The first two phases of the analysis have developed constraint-based descriptions of the system in terms of the functional capabilities of the systems (WDA) and in terms of the constraints affecting the activity (ConTA). It is possible to consider how these constraints affect the distribution of work and the allocation of function by using these descriptions as templates. Actors can be mapped onto these representations to show where they can have an influence on the system. This mapping allows the analyst to see a graphical summary of the constraints, dictating who has the capability of doing what. At this stage, the focus is entirely on capability, no judgement is made on which actor is best placed to perform a function. In this example, the key actors were identified by the SMEs as:

- CAOC / Fires – CAOC's (Combined Air Operations Centre) work at a tri-service level coordinating air operations and deconfliction in time and space. The CAOC produce the Air Tasking Order (ATO) and Airspace Control Orders (ACO).
- Aircrew – the aircrew fly the aircraft and are ultimately responsible for the planning and subsequent execution plan.
- Sqn MPS Operator – the MPS operator works more in a administrative role assisting the aircrew in creating plans and transferring data onto the MPS
- Ops officer / Commander – the ops officer is normally involved with the planning of future operations, the Commander is normally involved in current actions.
- EWO – the Electronic Warfare officer is a technical specialist in an advisory capacity; they provide information about enemy and friendly capabilities. Advice is also given on the best tactics to neutralise threats.

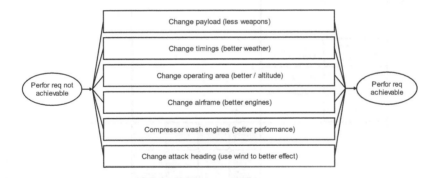

Figure 5.7   Example MPS Strategies Analysis

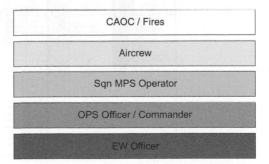

Figure 5.8   Colour key for actors in the domain

Figure 5.9 shows the ADS coded to indicate where each of the actors can influence the system. The coding is limited to the purpose related functions and the object-related process levels of the hierarchy. The higher levels, representing the functional purpose and values and priority measures, are considered applicable to all actors in the system, in the interest of clarity, these are not coded. This modification of the ADS provides a concise graphical summary, which forms the basis for the coding of the Contextual Activity Template (see Figure 5.10).

The Contextual Activity Template can be coded to show which actors can perform work functions in different situations (see Figure 5.10). Cells occupied by more than one actor indicate that activity can be supported by either or all of the identified actors. At this stage there is no consideration of which of the actors is best placed to conduct the activity, nor is there consideration of the best way

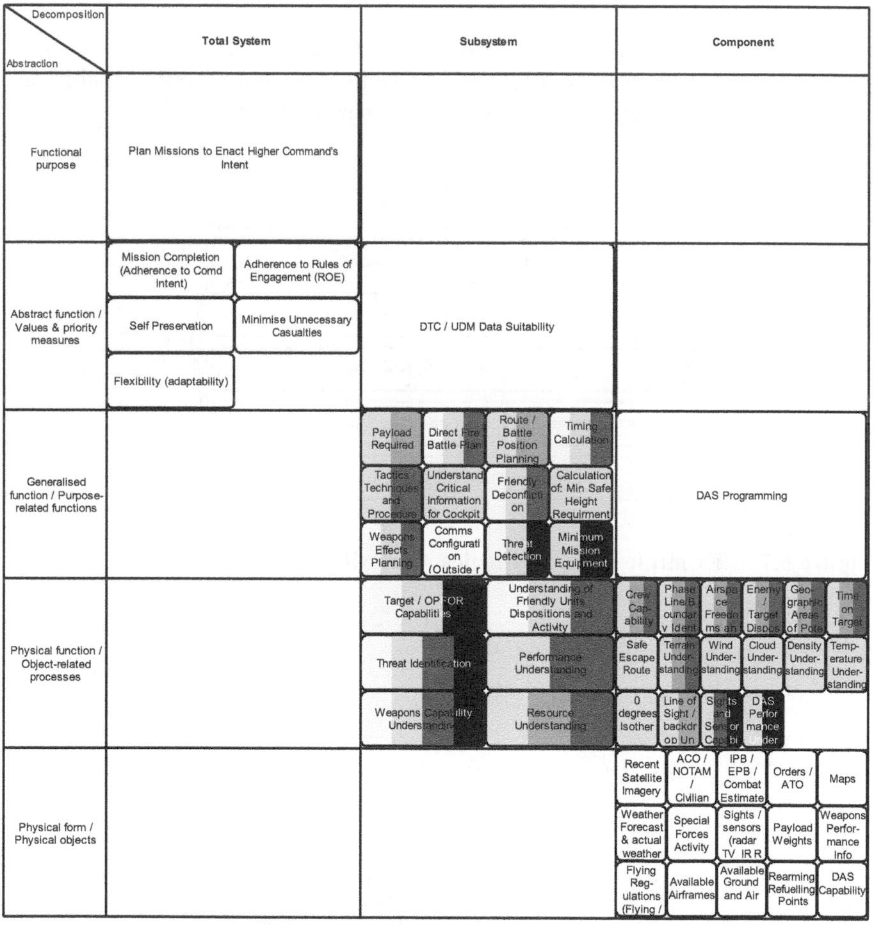

Decomposition / Abstraction	Total System	Subsystem	Component
Functional purpose	Plan Missions to Enact Higher Command's Intent		
Abstract function / Values & priority measures	Mission Completion (Adherence to Comd Intent); Adherence to Rules of Engagement (ROE); Self Preservation; Minimise Unnecessary Casualties; Flexibility (adaptability)	DTC / UDM Data Suitability	
Generalised function / Purpose-related functions		Payload Required; Direct Fire Battle Plan; Route / Battle Position Planning; Timing Calculation; Tactics Techniques and Procedure; Understand Critical Information for Cockpit; Friendly Deconfliction; Calculation of: Min Safe Height Requirement; Weapons Effects Planning; Comms Configuration (Outside r; Threat Detection; Minimum Mission Equipment	DAS Programming
Physical function / Object-related processes		Target / OP EOR Capabilities; Understanding of Friendly Units Dispositions and Activity; Threat Identification; Performance Understanding; Weapons Capability Understanding; Resource Understanding	Crew Capability / Ident; Phase Line/Boundary Terms an; Airspace Freedoms; Enemy Target; Geographic Areas of Pote; Time on Target; Safe Escape Route; Terrain Understanding; Wind Understanding; Cloud Understanding; Density Understanding; Temperature Understanding; 0 degrees Isotherm; Line of Sight / backdr op Un; Sights and Sensor Capabi; DAS Performance Under
Physical form / Physical objects			Recent Satellite Imagery; ACO / NOTAM / Civilian; IPB / EPB / Combat Estimate; Orders / ATO; Maps; Weather Forecast & actual weather; Special Forces Activity; Sights / sensors (radar TV IR R; Payload Weights; Weapons Performance Info; Flying Regulations (Flying /; Available Airframes; Available Ground and Air; Rearming Refuelling Points; DAS Capability

Figure 5.9   ADS coloured to show actors activity

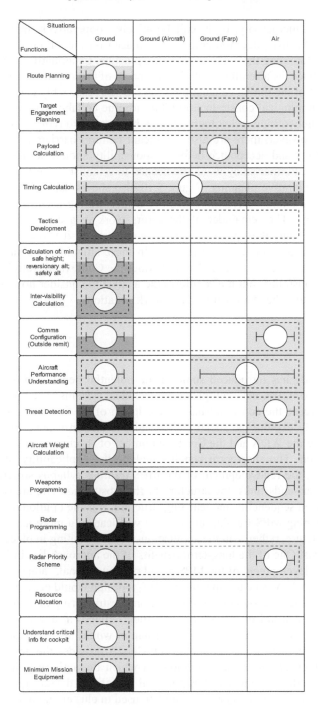

Figure 5.10   CAT coloured to show actors activity

of completing the activity; be it; collaboratively, cooperatively or by one actor in isolation. At this stage, the emphasis is placed on modelling the constraints rather than addressing the optimum working practice. Examination of Figure 5.10 reveals that once the aircraft has left the ground all of the identified activities (with the exception of the timing calculations) can only be reasonably conducted by the aircrew. Technological constraints prohibit airborne collaborative working. With advances in networking technologies, it may be possible to remove some of these constraints; however, there are also significant cultural barriers to be addressed relating to trust and acceptance before responsibility should be delegated away from the pilot. Further study would be needed to establish the effect of real-time airborne collaborative planning. Figure 5.10 also shows that whilst planning on the ground the aircrew have the capability of performing each of the identified work functions. The remainder of the actors work in a capacity to assist the pilots in developing their plans. Due to time constraints, it is often required that work functions are conducted in parallel. In these situations, collaborative and cooperative working is essential. Not only does the Contextual Activity Template capture the constraints but it also allows the analyst to consider how workload is distributed within the team within given work situations.

Conclusions

This case study has introduced some of the potential benefits of exploring the SOCA phase of CWA with complex sociotechnical systems. The approach taken has been to reuse the constraint-based description of the first two phases (WDA and ConTA) to explore the social and organisational constraints. In the process of conducting the WDA and the ConTA a number of short and long term benefits were extracted. The short-term benefits include the applicability of the WDA for informing the redevelopment of the MPS training syllabus structure. Based upon its means-ends links, the structure of the Abstraction Hierarchy forms the basis for lesson sequencing and teaching structure. It is the opinion of the authors that redeveloping MPS training in this way will lead to a more activity-focused teaching structure, rather than the application-focused training in place at the time of the analysis. One of the long-term benefits of the approach lies in its ability to guide future development of the MPS based on a functional, rather than physical interface. The CWA indicates that future MPS redesign would benefit significantly from task-orientated groupings of information. Restructuring the interface would provide users with all of the information they required at any one time within the same window. This grouping of information would also prompt the user to consider context specific information. The ConTA provides the developers with a greater understanding of the situations in which the activity is likely to take place. This understanding has the potential to inform the design of situation specific interfaces. These interfaces could be used to cluster and display pertinent information dependent on the current work situation.

The analysis of the MPS software tool revealed that it offers significant enhancements to the mission planning process. Planning with the MPS software can potentially be quicker, far more detailed and produce less planning errors, when compared to paper based planning. Further, the MPS software supports collaborative planning and automates many of the laborious and error prone components of the manual planning process. Despite this conclusion, our research also suggests that, although the MPS has the necessary functionality to support efficient mission planning, ultimately the design of the software's user-interface is under optimised. The design of the user interface, in the analysed system, makes it difficult for users to navigate to related data. This is predicted to have a negative impact on planning time, training time, user errors and frustration. The examined system represents one stage of the transition from an analogue to a fully network enabled system. From the analysis of the system, it is clear that there are technological constraints limiting the system flexibility, particularly within the distribution of tasks and the allocation of function. However, there also seem to be other factors preventing the system from fully exploiting the new technology capabilities. The MPS system appears to be little more than a digitisation of the analogue process, with activities in the digital system conducted in the same way as they were in the analogue system. An approach has been taken to automate 'mandraulic' processes in the planning activity; however, the system contains a significant amount of flexibility that has yet to be exploited. This potential flexibility could be explored to remove some of the constraints affecting the allocation of function.

The results from the SOCA phase (Figure 5.10) graphically show the distribution of activity between the actors within the system. It is clear that the aircrew are still responsible for the majority of the activity within the domain, particularly after the aircraft has taken off. As discussed, this is primarily due to technological constraints; however, the interface design has a significant role in supporting distributed working. The MPS software analysed does not actively support collaborative working. The analysis in Figure 5.10 clearly shows that many of the activities required in the first work situation (on the ground using MPS terminals) can be conducted by a range of different actor groups. The analysis has highlighted that, through the addition of data sharing protocols and a simple local area network, many of the less safety critical components of the task could be shifted away from the pilots to other actors in the domain. This could potentially expedite the planning process significantly. Stanton et al (2006) found that to exploit the benefits of distributed planning activities within complex systems fully; there is a need for compatibility in situation awareness. A networked system presenting a 'common picture' could assist in the development of this shared situation awareness. The framework presented also forms a basis for further exploration of work allocation. Whilst the approach discussed in this case study has concentrated on the modelling of constraints, it is contended that this representation forms a basis for exploring, in detail, the allocation of function between actors within each cell. From the developed description of constraints, potential combinations of working practices can be identified and evaluated to determine optimal practices.

When the formative systems approach used in the SOCA is compared to more 'traditional' normative approaches (EAST; Walker 'Walker, Guy' et al 2006a), it is clear that normative methods provide a much better basis for after action review. These normative approaches are therefore more suitable for diagnosing what has and should have happened, rather than predicting or postulating what can happen. The strength of CWA is that it provides an externally observable, constraints based, description of the world. Although methods such as EAST may fall short in describing formative behaviour, arguably they provide deeper systems based descriptions of cognition. It is for this reason that these methodologies are complimentary for fulfilling the aim of modelling complex sociotechnical systems.

Social Network Analysis (SNA) tells us where links exist between agents. In many cases (Houghton et al, 2006) the importance of these interactions is then derived from their frequency. SOCA on the other hand explains the constraints limiting the allocation of activity between the actor groups. By using the later phases of CWA to focus on the alternative strategies and system configurations, redundancy can be identified. The level of redundancy available informs the importance of a link. Without any redundancy, a link is pertinent; however, an important link identified by SNA may not be required within a system if there is another way of achieving the same end state. This more formative approach therefore compliments SNA by providing a validation of the statistical metrics based upon frequency of use.

The recommendations in this case-study (along with other findings in the wider HFI DTC project) were passed on to the developers of the next generation mission planning tool and have been considered in the design and development of a new system.

Evaluation – Using Work Domain Analysis to Evaluate the Impact of Digitisation on Command and Control

This section introduces a new approach for evaluating the impact of technological change on complex sociotechnical systems. The approach uses Work Domain Analysis as a theoretical base for extracting the key factors that influence system performance. The process has been designed to be expeditious, in terms of both construction and data collection. The approach uses the opinion of subject matter experts to evaluate the impact of each of the Abstraction Hierarchy nodes on system performance. This approach was used to evaluate the effects of digitisation within land based military head quarters, at brigade and battlegroup levels. The proposed approach proved sensitive enough to reveal significant differences between the old and new systems. The description of the same system at a number of levels of abstraction allows the analyst to develop a high level rating of the system as well as understanding of the key factor that have influenced this opinion.

The selection of performance measures for the evaluation of complex sociotechnical military domains is an inherently difficult process. These domains are frequently influenced by a wide range of often-conflicting factors. These factors each have the potential to affect the overall performance of the system in unexpected ways. In practice, it is extremely difficult and time consuming to collect data on the performance of each of these factors. Further challenges exist in the synthesis of this data and the determination of its relative importance.

According to Yu et al (2002) traditional measures of task performance (e.g. task completion time) are objective, but frequently do not have a compelling theoretical basis and are not sensitive enough to reveal differences between experimental groups. Other measures (e.g. eye movements, verbal protocols) can provide greater scientific insight, but frequently suffer from being extremely time-consuming or subjective to analyse. As with many domains, there is a plethora of possible measures to take. According to Crone et al (2007), the selection of these measures via task-based methods is influenced by several factors such as practitioner experience (Muckler and Seven, 1992), resources available, and criteria including reliability, validity and sensitivity. Traditional measures such as time taken to complete a task are not always appropriate as time scales are often governed by external environmental factors. In military systems the planning time available is often fixed and therefore, unavailable as a measure of system performance.

An evaluation of the impact of digitisation upon the command and control process was required. The domain investigated was a land based military battlefield environment. The system was evaluated at the Brigade (Bde) and the Battlegroup (BG) headquarters (HQ) levels of command. To gain an understanding of the level of system improvement the traditional analogue process was used as a benchmark. This analogue system was to be compared with its new digital counterpart. The digital system introduces features such as secure encrypted radio and computer terminals into the system. These terminals display geospatial representations of the environment and support unit tracking and graphical overlay production. These networked terminals also support data sharing, and the creation and transmission of digital versions of planning products such as orders.

Traditional approaches to task analysis, such as Hierarchical Task Analysis (Shepherd, 1985; Stanton, 2006) are based upon a reductionist approach. They decompose activity into its constituent parts and use these to measure the performance of each sub-activity. This can then be used as an effective tool to compare similar competing system to reveal differences in efficiencies. However, the introduction of new technologies, in this case digitisation, brings with it new, previously unavailable capabilities; these capabilities allow work to be conducted in revolutionary ways. Crone et al (2003) point out that when a new system is introduced to perform the same operational task as the systems it is replacing, the same task based performance may be used to assess the effectiveness of the new system. However, the applicability of these measures may change if the new system provides radically different functionality and there is no knowledge about the functionality and its implications for human-system integration. As the system under examination provides new functionality such as data transmission and new levels of collaborative distributed working, a direct comparison, based upon task analysis, is likely to reveal distorted results.

The data collection exercise took place over a space of three weeks. This large-scale trial involved a fully functional division, Brigade and Battlegroup. Orders were composed at division and sent to Brigade, who in turn created orders for Battlegroup. Battlegroup then prepared and distributed orders to each of the companies under its command. The trial was set up to represent an operational situation closely. As with many field trials, access to data and subject matter experts was limited. An approach was therefore sought that could provide bespoke, theoretically grounded, contextually applicable measures of task performance. These were also required to be generated and applied in an expeditious manner. The developed approach is not intended to be used in isolation; rather, it is intended as a complementary approach to objective measures of performance. The approach uses subject matter expert opinion to apply a rating to each of the derived measures.

The measures used are derived from the Abstraction Hierarchy (AH; Rasmussen 1985); part of CWA framework. CWA is a technique used to model complex sociotechnical systems. The use of the framework to model military command and control is not a new concept, it has been successfully applied in the past (e.g. Burns et al, 2003; Chin et al, 1999; Cummings and Guerlain, 2003; Jenkins et al, in 2008a; Jenkins et al, in 2008b; Lamoureux et al, 2006; Lintern et al, 2004; and Naikar and Saunders, 2003). The framework is used to model different types of constraints, when used in its entirety it builds a model of how work can proceed within a given system. According to Vicente (1999), there are five separate phases, each phase considering different constraint sets. The analysis used in this chapter focuses on the initial phase; Work Domain Analysis (WDA). WDA provides a description of the domain or environment in which activity can take place. By focusing on constraints, an emphasis is placed on what is possible rather than what does, or should happen. The model is independent of actors and of activity. According to Crone et al (2007), the CWA framework is important for several reasons. First, it provides a way to categorise measures into meaningful groups. Second, it provides a mechanism to explore relationships between system components. The main tool used within WDA is the Abstraction Hierarchy or Abstraction-Decomposition Space. The approach models the system at a range of levels of abstraction (typically five). At the highest level the overall system's aims and objectives are considered. At the lowest level, the individual components that make up the system are considered. Through a series of structural means-ends links, indicating affordance, it is possible to model how individual components can have an influencing effect on the overall system purpose. In this case, a model has been generated that captures the overall purposes of battlefield management and planning. The upper four levels of the Abstraction Hierarchy are independent of technology; they concentrate on the overall objectives of the system and the means by which they can be influenced. It is only at the very bottom of the hierarchy where the individual components and technologies required are considered. Therefore, the uppermost four levels of the model are technologically agnostic and therefore applicable to both the analogue and the digital process. As the model is applicable for both the new and the benchmark system, it serves as an ideal mechanism for comparing how well the new digital system supports the command process.

The Abstraction Hierarchy model produces descriptions of the same system at a number of levels of abstraction. At the base of the hierarchy the physical functions of the system components can be examined, these functions can then be evaluated in terms of their contribution to the purpose related functions, in turn these can be compared to investigate their effect on the values and priority measures and finally on the overall functional purpose. By evaluating the performance of each part of the system it is possible to understand which of the physical functions have a significant impact on the overall functional purpose. In this case, a subject matter expert is used to evaluate the impact of each node in the hierarchy: each of the nodes within the Abstraction Hierarchy is used to ask the question – is this part of the system significantly better, about the same, or significantly worse than the benchmark?

This approach therefore, stands alone as a theoretically grounded technique for developing and testing measures of performance. Its theoretical grounding creates a clear and rational means for selecting performance measure. It also provides a clear audit trail capturing a record of how the measures are derived. By combining this approach with traditional empirical approaches it is possible to provide further evidence to substantiate the findings of this approach, further the proposed approach adds context and provides important information on the relationship between performance measures.

Method

An Abstraction Hierarchy was constructed for the command process (see Figure 5.11). The document was created using a number of resources: A guide relating to the planning process trained to staff officers (the Combat Estimate Process; CAST, 2007) provided a clear description of the required information and products of each stage of the planning cycle. Operational understanding was provided by a training video (BDFL, 2001) of a Battlegroup in a 'quick attack'. This was supplemented with the experience of the author in military and non-military command and control domains. The completed model was then validated with a military subject matter expert. As previously stated with the exception of the bottom level (physical objects), the model is technologically agnostic. The model is also non-specific to any level of command; it represents the generic command and control process and is therefore applicable for both Bde and BG in the analogue or the digital process.

At the highest level of abstraction, the overall functional purposes of the command and control system are listed. These capture the overall reason that the system exists. These purposes are independent of time, they exist for as long as the system exists. In this case, the overall purposes have been listed as 'Battlefield

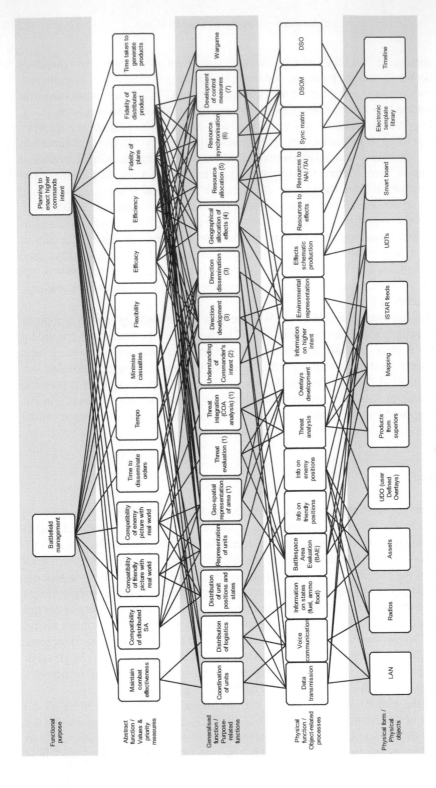

Figure 5.11 Abstraction Hierarchy for the command process

Functional purpose

Abstract function / Values & priority measures

Generalised function / Purpose-related functions

Physical function / Object-related processes

Physical form / Physical objects

Battlefield management

Planning to enact higher commands intent

Maintain combat effectiveness

Compatibility of distributed SA

Compatibility of friendly picture with real world

Compatibility of enemy picture with real world

Time to disseminate orders

Tempo

Minimise casualties

Flexibility

Efficacy

Efficiency

Fidelity of plans

Fidelity of distributed product

Time taken to generate products

Coordination of units

Distribution of logistics

Distribution of unit positions and states

Representation of units

Geo-spatial representation of area (1)

Threat evaluation (1)

Threat integration (COA analysis) (1)

Understanding of Commander's intent (2)

Direction development (3)

Direction dissemination (3)

Geographical allocation of effects (4)

Resource allocation (5)

Resource synchronisation (6)

Development of control measures (7)

Wargame

Data transmission

Voice communication

Information on states (fuel, ammo food)

Battlespace Area Evaluation (BAE)

Info on friendly positions

Info on enemy positions

Threat analysis

Overlays development

Information on higher intent

Environmental representation

Effects schematic production

Resources to effects

Resources to NAI / TAI

Sync matrix

DSOM

DSO

LAN

Radios

Assets

UDO (user Defined Overlays)

Products from superiors

Mapping

ISTAR feeds

UDTs

Smart board

Electronic template library

Timeline

management' and 'Planning to enact higher commands intent'. In this model, these are considered the sole reasons for the Command Centre's existence.

At the second level down; the values and priority measures, a number of metrics are captured for evaluating how well the functional purposes are being performed. The links up to the functional purposes indicate which of the measures relate to each functional purpose. In this case the majority of measures apply to both functional purposes with the exception of; Fidelity of plans; Fidelity of distributed product; and Time taken to generate products which relate solely to the functional purpose of planning.

In the middle of the hierarchy, the purpose related functions are listed. These are the functions that can be conducted, that have the ability to influence one or more of the values and priority measures. The numbers in brackets in some of the nodes relate to the number of the stage in the combat estimate process.

The second level from the bottom, the object related processes captures the processes that are required to be conducted by the physical objects in order to perform purpose related functions. An example of this is voice communication; voice communication allows the system to perform Coordination of units; Distribution of logistics; Distribution of unit positions and states; and Direction dissemination. It should be noted that the model represents what can happen rather than what does or should, the emphasis with this approach is on constraints. At the object related purposes level, the model concentrates on what the objects can physically do independent of the overall system function purposes.

The bottom layer of the model lists the physical objects contained within the system. This is the only part of the system that would differ for the analogue and the digital system. The discussed model, therefore, provides a description of the system's constraints and capabilities. The nodes are connected by a series of means-ends links. Any node can be taken to form the question of what is required, the linked nodes above indicate why this required, the linked nodes below answer the question of how this can be achieved.

In total eleven subject matter experts (five from BG and six from Bde) were surveyed at the end of the three week trial to establish their opinion of the new digital system. The subject matter experts comprised of the senior officers from each of the cells of the command team. The model was briefly explained to the participants as a technologically independent system description. Participants were asked to consider how the new technology had influenced their role and summarise if the system is; significantly better, significantly worse, about the same or if the new system does not support the process. The instructions read to the participants are transcribed below:

'Would it be possible to take 10 minutes of your time to ask you a few questions?

From the literature and from our observations over the past few days we have developed a model that we feel represents the activity you as a head-quarters are trying to conduct. The model we have developed is independent of technology, such that, it applies to both the old analogue paper-based process and the new digitised system you have been using that is supported by the digital system. The model starts at the bottom describing the key processes and assets within the system. At the next level up the model describes what these assets allow you to do, for example answering the seven questions posed in the 'Combat estimate'. At the next level up we have captured what we consider to be the key metrics that will determine how well the overall head-quarters system is performing. Finally, at the top we look at the overall reasons why the headquarters system exists; to plan and manage battles.

What I would like to do is go through these, one at a time, and get your opinion of how the introduction of the digital system under trial has affected your role. I would like you to summarise and say if it is; significantly better, significantly worse, about the same or if the new system does not support it.

If you feel unable to comment on any of the questions because it does not impact your role let me know and we can leave it blank'.

Participants were then read each of the lines in Figure 5.12 (derived from the Abstraction Hierarchy nodes) from the bottom up (with the exception of the purpose related functions section which was read in temporal order, from top to bottom) and asked to rate them. The results can be seen on the right hand side of Figure 5.12. A plus indicates that the participant rated the system as significantly better, a minus that they rated it as insignificantly worse, an equals indicates they rated it as about the same, an 'N' indicates that they believed the new digital system did not support this function, and a blank cell indicates that the respondent did not feel qualified to rate that part of the system.

Results

In summary there seemed to be little evidence to suggest that the proposed digital system supports the planning process. The overall opinion of the users surveyed is that the digital system observed is detrimental to the planning process. This opinion can be explained at the purpose related functions level, the processes outlined in the combat estimate planning process seemed unsupported by the software. The digital version seems to offer little or no additional functionality and as the object related process level shows the products take longer to generate as a direct result of user interaction problems.

The ratings for the battlefield management aspect of the systems seem far more positive; however, the results would suggest that this perceived benefit seems to be almost solely related to the addition of the secure-voice radio. The collected data would suggest that the computer terminal part of the system, in its current states, provides relatively little to enhance battle management.

In a number of the questions there seemed to be a significant disagreement between the responses at BG and at the Bde level. Bde seemed more likely to

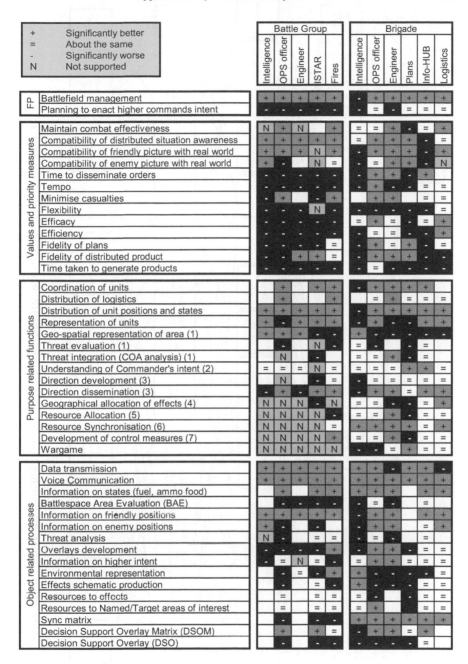

Legend:

+	Significantly better
=	About the same
-	Significantly worse
N	Not supported

		Battle Group					Brigade					
		Intelligence	OPS officer	Engineer	ISTAR	Fires	Intelligence	OPS officer	Engineer	Plans	Info-HUB	Logistics
FP	Battlefield management	+	+	+	+	+	-	+	+	+	+	+
FP	Planning to enact higher commands intent	-	-	-	-	-	-	=	-	=	=	=
Values and priority measures	Maintain combat effectiveness	N	+	N		+	=	=	+	-	=	+
	Compatibility of distributed situation awareness	+	+	+	+	+	=	+	+	+	-	=
	Compatibility of friendly picture with real world	+	+	+	N	+	-	+	+	+	-	+
	Compatibility of enemy picture with real world	+	-		N	=	-	=	+	+	-	N
	Time to disseminate orders	-	-	-	-	-	-	+	+	-	+	
	Tempo	-	-	-	-	-	-	+		-	=	=
	Minimise casualties	-	+		-	+	-	+	+		=	=
	Flexibility				-	N	-	-	-	-	-	=
	Efficacy	-	-	-	-		=	=	=	-	-	+
	Efficiency	-	-	-	-		-	=	=	-	-	
	Fidelity of plans	-	-	-	-	=	-	+	=	+	-	=
	Fidelity of distributed product	-	-	+	+	=	-	+	+	+	-	-
	Time taken to generate products	-	-	-	-	-	-	=	-	-	-	-
Purpose related functions	Coordination of units		+		+	+	-	+	+	+	+	
	Distribution of logistics		+			+	=	+	=	=	=	
	Distribution of unit positions and states	+	+	+	+	+	-	+	+	+	+	+
	Representation of units	+	-	+	+	+	-	+			+	+
	Geo-spatial representation of area (1)	+	+	+	-		+	-	-	-	-	-
	Threat evaluation (1)		-		N	-	=	-			=	
	Threat integration (COA analysis) (1)		N		-		=	=	+	-	=	
	Understanding of Commander's intent (2)	=	=	=	N	=	=	=	=	+	+	=
	Direction development (3)		N		-	=	-	=	=	=	=	=
	Direction dissemination (3)	-	+		+	+	-	+	=	+	+	+
	Geographical allocation of effects (4)	N	N	N	-	N	=	=	-	-	=	+
	Resource Allocation (5)	N	N	N	N	-	=	=	+	-	=	=
	Resource Synchronisation (6)	N	N	N	N	=	+	+	+	+	=	+
	Development of control measures (7)	N	N	N	N	+	=	=	+	-	=	=
	Wargame	N	N	N	N	N	-	-	=	+	=	=
Object related processes	Data transmission	+	+	+	+	+	+	+	-	+	+	-
	Voice Communication	+	+	+	+	+	+	+	+	+	+	+
	Information on states (fuel, ammo food)		+		+	+	+	+	+		+	+
	Battlespace Area Evaluation (BAE)		-	-	-	-	-	=	-		=	
	Information on friendly positions	+	+	+	+	+	-	+	+		+	+
	Information on enemy positions	+			-		-	+	+		=	+
	Threat analysis	N	-		=	=	-	=	-		=	
	Overlays development	-	-	-	-	+	-	+	+	-	=	=
	Information on higher intent	-	=	N	=	-	=	+	+	=	=	=
	Environmental representation		=		-	+	+	-	-		-	=
	Effects schematic production			=		+	+	-	-		=	=
	Resources to effects		=		=	=	=	-			=	=
	Resources to Named/Target areas of interest		=		=	=	=	+			=	=
	Sync matrix		-		-	-	+	+	+	+	+	+
	Decision Support Overlay Matrix (DSOM)		+		+	=	-	+	+	=	+	
	Decision Support Overlay (DSO)						-	-	-	-	=	

Figure 5.12 **Subjective opinion of the digital system in work domain terms (blank cells indicate that the respondent did not feel qualified to comment)**

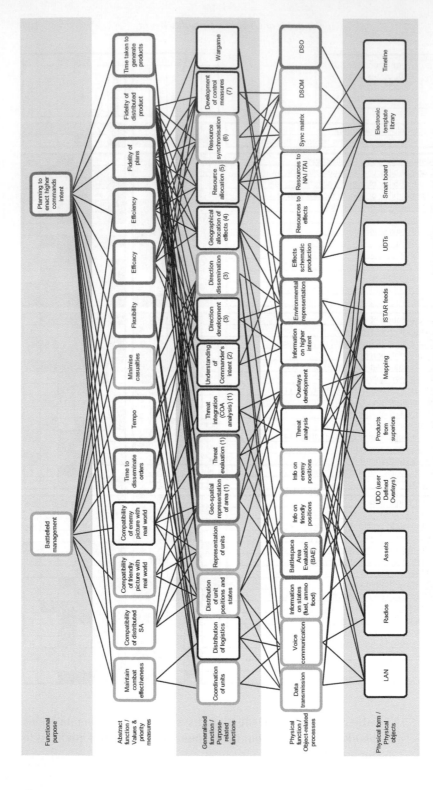

Figure 5.13 Abstraction Hierarchy showing summary of cells significantly worse and significantly better

express a positive opinion about the way that the system supported the planning process. The results of this approach would therefore suggest that the system is better suited to command at a higher level. Further investigation of the system at a division level would be required to reinforce this hypothesis.

Conclusions

This section has introduced a new structured approach to the selection of measures of performance. The technique uses Work Domain Analysis to describe the systems at a number of levels of abstraction; these differing perspectives of the same system provide an overall rating of system performance as well as an indication as to which of the constituent parts of the systems cause these effects. The approach provided has proved sensitive enough to reveal significant differences between the two systems (analogue and digital).

Whilst it is difficult to draw conclusions from a limited trial; not subject to tight experimental controls. The discrepancies between the different levels of command would appear to provide some evidence for the requirement of further customisation within the system. Both BG and Bde plan to the same planning combat estimate; however, it should be noted that this is an extremely flexible approach. The more rigid prescriptive working practices encouraged by the digital system would appear to be incompatible with this. The comparison has revealed that the digital system does support battlefield management; however, by viewing the system at a number of levels of abstraction it is possible to reveal that this is predominantly due to the introduction of secure encrypted radio. The approach is also able to inform decisions as to where design improvements may influence system design and where functionality should be excluded from the digital system.

As with all subjective assessment, this approach is open to participant biases; however, when used in conjunction with objective measures this approach serves as a powerful tool for the evaluation of complex sociotechnical systems in military command and control. Although untested beyond this domain it is postulated that due to the strong theoretical basis this approach for evaluation would be applicable for any system suitable for analysis by the CWA framework.

Chapter Summary

This chapter has addressed how CWA can be used for the purposes of analysis, informing design, and evaluation. The approach has been applied to two examples, a Helicopter Mission Planning System, and a battlefield command and control system. As described in the introduction to this chapter, both of these domains fall under the category of complex sociotechnical systems. Whilst, the approach has merit in the analysis of simple artefacts, it is through the exploration of the often-conflicting constraints of these complex systems that the real benefits of a systems based approach, such as CWA, become clear. In the example of the MPS, the analysis as extended to include later phases of the framework, this more comprehensive model of the system is used to develop broad design recommendations. A new evaluation system was proposed for the battlefield command and control system, this approach successfully uses the descriptions of the system in the Abstraction Hierarchy to compare two systems directly.

Both of these case studies provide descriptions of the analysis undertaken. The first case study also identifies opportunities for design improvements, the detailed look at the SOCA phase of the analysis also encourages the analyst and the designer to consider how the implications of; location, time and actor can influence the design of the system. The SOCA phase raises questions about the correct mix of cooperation and collaboration as well the suitability of distributed working practices.

The second case study provides further description of the WDA phase in a new domain. The case study also introduces the concept of using the Abstraction Hierarchy to generate measures of performance; these measures can be used in a survey of subject matter experts. The approach described uses CWA to create a theoretically grounded evaluation method that can be rapidly developed, administered and evaluated to develop an understanding of system performance at a number of levels of abstraction. Whilst this method builds upon the work of Crone et al (2003), it applies the measures in a unique manner. The approach lacks objective ratings of each measure; however, it captures user opinion and can be rapidly applied.

This chapter has shown that as the complexity of a domain increases, so does the benefit, and indeed requirement, for a constraint based approach such as CWA. Although many benefits can be seen from this approach it is not until we explicitly consider the design of these systems that the true benefit of CWA is unearthed.

Chapter 6
Using CWA to Design for Dynamic Allocation of Function

Chapter Introduction

The objective of this chapter is two fold, firstly it reinforces the benefits of applying CWA beyond the initial phases, illustrating the relevant interconnections; secondly the analysis shows how design recommendation can be extracted from the analysis products. A clear process is shown of how a number of interfaces were developed as a direct result of the CWA.

The case study uses a command and control micro-world example to illustrate how each of the five phases can be used to describe the constraints within the micro-world domain from a different perspective. Using a number of the CWA phases bespoke interfaces are developed for this micro-world; the Social Organisation and Cooperation Analysis phase is used to develop design requirements for role specific customisable interfaces. These interfaces have been specifically developed to communicate real time reconfiguration of the network through each of the individual interfaces; the reallocations of functions or roles are communicated to the actors through changes to the interface.

Whilst the first half of this chapter focuses heavily on the analysis, the second half of this chapter attempts to make the link between analysis and design more explicit. Two types of interface are devoted; functional and physical. The physical interfaces are developed for actors interacting closely with the interface at lower levels of abstraction, whereas, the functional interfaces are designed for those less well coupled with the environment at a system management level, these actors work at high levels of abstraction.

CWA of the Sensor to Effecter Paradigm

This paradigm has been developed to conduct human factors research into rapidly reconfigurable networks. The network facilitates the exchange of information between agents in the field and a series of centrally located commanders. The environment developed allows the manipulation of dependent variables to establish the most efficient network structure for a variety of different scenarios. Cognitive Work Analysis has been used to analyse and model the experimental system, establishing a model that can be used to hypothesise the implications of changes to the network structure and the resulting influence this will have on the system and the agents contained within. The analysis uses a Work Domain Analysis to capture the purpose, capabilities and constraints of the system. A Control Task Analysis outlines the task required to fulfil the purpose of the system. This task is broken down in a Strategies Analysis, which explains the possible ways that the system can be configured to enable the same end state. A Social Organisation and Cooperation Analysis elucidates which of the actors within the system can perform the tasks required. Finally a Worker Competencies Analysis describes the resulting behavioural characteristics the actors will exert depending on the level of tasks they are assigned. In conclusion, example network configurations for a number of conceived scenarios are presented.

The Domain

The system under consideration is based upon a micro-world paradigm developed as a Human Factors test bed to conduct research into military command and control. The paradigm has been developed to represent a range of command and control domains (both military and non-military). The environment consists of an urban setting of approximately 20 hectares. Within the environment, there are a number of concealed 'targets', requiring the system's attention. There are two types of actor distributed within the environment. The first type are reconnaissance units known as 'sensors'. Sensors have the ability to sweep a geographic area and identify targets that need to be attended to. The second type are effecters who are responsible for attending to identified targets. In this simple paradigm, sensors are the only actors that can detect targets and effecters are the only actors who can attend to previously identified targets.

The system also consists of two levels of command known as Gold and Silver, one silver commander is in charge of sensors and one the effecters. The Gold commander has the responsibility for the entire system.

Dependent upon the way that the system is configured, there are a number of ways that information can be transmitted between the sensors and effecters. The system can be set up to mimic a traditional hierarchy (see solid lines in Figure 6.1) enabling information to be sent via the commanders with information travelling up the hierarchy and then back down to the units in the field; alternatively information can be sent directly from peer-to-peer (see dotted lines in Figure 6.1).

When sending peer-to-peer, the network can be configured so that a sensor can be linked to an effecter (Figure 6.1). Alternatively, the system can be configured so that the sensor has the ability to select the recipient of the information (See Figure 6.2).

The system is significantly reconfigurable; the exact configuration choice will be influenced by a number of variables. These include:

- Number of units – how many sensors
- Ratio of effecters to sensors – how many effecters per sensor

- Ratio of targets to sensors – how many targets per sensor
- Complexity of task – are there a number of conflicting requirements
- Complexity of the target – is interpretation of the target required
- Ambiguity of information – is it clear what the information represents
- Type of information transmitted (e.g. data, voice)

Whilst it is accepted that this model is a simplified account of sensor to effecter networks found in operational environments, the model does attempt to capture the essential features. Other sensor to effecter network analysis have used similar paradigms with some success (e.g. Dekker, 2003)

The Analysis

Much of the current literature focuses heavily on the first phase of CWA, work domain analysis (WDA). As Hajdukiewicz and Vicente (2004) point out, WDA does not explicitly deal with any particular worker, automation, event, task,

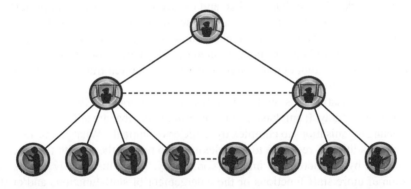

Figure 6.1   Diagram showing possible structures of sensor and effecters

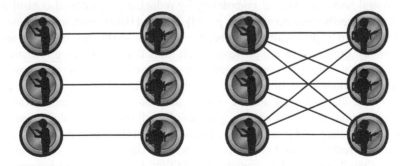

Figure 6.2   Diagram showing different peer-to-peer relationships

goal or interface, because of this, its use in a practical setting is often called into question (Darses 2001). Hajdukiewicz and Vicente (2004) attempt to resolve this issue by making the relationship between WDA and Task Analysis more explicit. This chapter aims to build upon this idea of making the links between phases explicit extending this approach to each of the five phases described by Vicente (1999). The chapter attempts to expose the benefits of the lesser discussed latter three phases of CWA using the fourth phase SOCA to inform the design of rapidly reconfigurable interfaces.

According to Naikar (2006), Cognitive Work Analysis (CWA) is continuing to gain momentum 'as an approach for the analysis, design, and evaluation of complex sociotechnical systems'. A sociotechnical system is a specific expression of sociotechnical theory. Sociotechnical theory is founded on two main principles. One is that the interaction of social and technical factors creates the conditions for successful (or unsuccessful) organisational performance. This interaction is comprised partly of linear 'cause and effect' relationships (the relationships that are normally 'designed') and partly from 'non-linear', complex, even unpredictable relationships (the good or bad relationships that are often unexpected). Whether designed or not, both types of interaction occur when socio and technical elements are put to work. The corollary of this, and the second of the two main principles, is that optimisation of each aspect alone (socio or technical) tends to increase not only the quantity of unpredictable, 'un-designed' relationships, but those relationships that are injurious to the system's performance. Sociotechnical theory, therefore, is about 'joint optimisation'. CWA is a structured framework that helps to achieve this state.

Joint optimisation was analysed in Trist and Bamforth's (1951) now classic sociotechnical study. It emphasises a number of factors that are still relevant today for relating organisations to complex dynamic environments. What can be referred to as the 'traditional' response to complexity 'is to restore the fit with the external complexity by an increasing internal complexity. [...] This usually means the creation of more staff functions or the enlargement of staff-functions and/or the investment in vertical information systems' (Sitter, Hertog and Dankbaar, 1997, p. 498). On the other hand, from a sociotechnical perspective, '...the organisation tries to deal with the external complexity by reducing [...] internal control and coordination needs' (Sitter et al., 1997, p. 498). Trist and Bamforth described how the latter case could be practically achieved. They emphasised the role of small 'semi-autonomous' groups, who are presented with tasks that are 'minimally critically specified'. In other words, simple organisations with complex jobs (as opposed to the more traditional case of complex organisations featuring simple jobs; Sitter, et al., 1997). Through these means, it is argued that supervision and leadership becomes internal to groups so that goals can be adjusted and problems solved. There is immediacy and proximity of trusted team members and opportunities for 'continuous, redundant and recursive interactions' (De Carvalho, 2006) and, as a result, shared situational awareness. In addition, the small group organisation design reduces interdependence so that local problems do not propagate through a larger social and organisational space. Finally, 'for each participant the task has total significance and dynamic closure' (Trist and Bamforth, 1951, p. 6). These are valuable attributes for contemporary military operations and circumscribe the various promises of Network Enabled Capability (NEC).

CWA provides a structured framework for sociotechnical design and, by extension, for realising the promises of NEC. SOCA (the fourth of CWA's five phases), recognises that organisational structures in many systems are generated on line and in real time by multiple, cooperating actors responding to the local context (e.g. Beuscart, 2005). In the words of sociotechnical theory this would be a demonstration of the autonomy granted to groups and the freedom members of a group have to regulate their own internal states and relate themselves to the wider system. SOCA is therefore, expressive of the 'simple organisation/complex job' philosophy. It is not necessarily concerned with planning upfront the nature of organisational structures that should be adopted in different situations. It is instead concerned with identifying the set of possibilities for work allocation, distribution and social organisation. SOCA explicitly aims to support flexibility and adaptation in organisations (the general systems principle of 'equifinality'; Bertalanffy, 1950) by developing designs that are tailored to the requirements of the various possibilities (the sociotechnical principle of 'multifunctionality; Cherns, 1987). Ironically, SOCA is one of the more neglected phases of CWA (most emphasis being given to Work Domain Analysis). The aim of this chapter is to redress this imbalance and to demonstrate the value of SOCA in relation to simple case study of organisational and interface design.

Work Domain Analysis

The first phase of CWA; Work Domain Analysis (WDA) is used to describe the domain in which the activity takes place independent of any goals or activities. Hajdukiewicz and Vicente (2004) are keen to point out that WDA does not explicitly deal with any particular worker, automation, event, task, goal or interface. Via a hierarchy, WDA captures the relationships between the physical objects and the system's overall purpose.

The first stage of this process is to construct an Abstraction Hierarchy (AH) of the domain (see Figure 6.4). The AH represents the system domain at a number of levels; at the highest level, the AH captures the system's raison d'être; at the lowest level the AH captures the physical objects within the system. In this simple sensor-effecter paradigm, the sole reason that the system exists is to detect and attend to targets within a predefined area. The system is evaluated against its ability to enact its purpose. This can be measured by a number of criteria, including: the time it takes the effecter to receive a target, how quickly all of the targets can be attended to (this could be achieved by attending to them based on the target's geographical position), the speed at which threat is reduced (this could be achieved by attending to the most dangerous targets first), and the number of errors made. In many circumstances, these criteria may be conflicting. An example of this conflict includes units approaching targets in threat priority order (left-hand image in Figure 6.3). If the same effecter prioritised the targets by their geographical position, the targets would be approached in a different order (right-hand image in Figure 6.3); whilst the route would be shorter and therefore faster to complete, the target with the greatest threat may be the last to be attended to. It could also be argued that speed to complete and error rates are conflicting constraints, the assumption being that careful, time consuming, planning reduces errors.

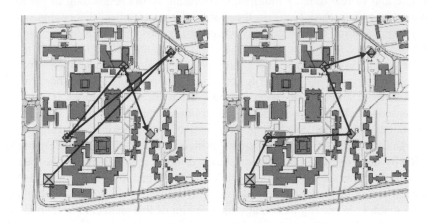

Figure 6.3   Example of conflicting requirements; by threat (on left-hand side) and by geographic location (on right-hand side)

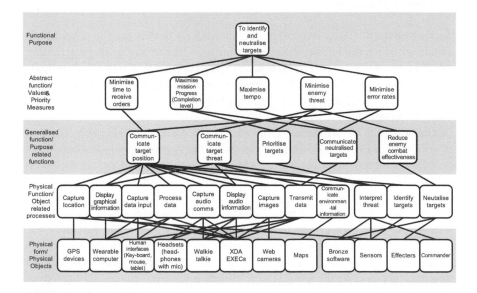

Figure 6.4   Abstraction Hierarchy for sensor-effecter activity

The bottom level of the AH shows each of the physical objects within the domain; in this case the nodes comprise all of the equipment within the domain. The level above this describes the functions that each of the objects can afford independent of the overall system purpose; in many cases, an object may perform a number of functions, in the same way a particular function may be afforded by a number of objects.

The purpose related functions in the middle of the AH are the functions required to perform the purposes of the system. Each of these levels can be linked by means-ends relationships using the why-what-how relationship. Any node in the AH can be taken to answer that question of 'what' it does. The node is then linked to all of the nodes in the level directly above to answer the question 'why' it is needed. It is then linked to all of the nodes in the level directly below that answer the question 'how' this can be achieved. Taking the example of the 'Capture Location' node in the physical function level, the question of 'what' is clearly answered. The node is then linked to nodes in the level directly above to answer the question of 'why' capture location, (in this case to calculate target position, and prioritise target). The node is then linked below to answer the question, 'how' can the location be captured, (in this case by GPS devices and by sensors).

The AH can then be decomposed based on levels of resolution through the system. In this case the system was decomposed using the following three categories of resolution: system, subsystem and individual components. Once decomposed, the data can be plotted on the Abstraction-Decomposition Space (ADS). In many cases, the nodes are decomposed along the diagonal of the ADS; moving across the decomposition axis as they move down the abstraction axis (see Figure 6.5). According to Hajdukiewicz and Vicente (2004), at higher levels of abstraction, participants tend to think of the work domain at a coarse level of resolution, whereas, at lower levels of abstraction, participants tend to think of the work domain at a detailed level of resolution. The functional purpose(s) of the system in most cases will apply to the total system; likewise, the individual physical forms are likely to be either components or subcomponents.

The application of WDA leads the analysts to focus on the reason the system exists considering each of the physical components against their ability to support this. The means-ends links within the Abstraction Hierarchy capture the flexibility of the system illustrating that the system can often be configured in a number of ways to achieve the same end state. This formative event, independent understanding of the constraints within the system, forms a basis for further examination of the work domain. There is also great benefit in representing the same domain at a number of levels of abstraction. Those tightly coupled with the environment (sensors and effecters) are required to predominantly consider the domain in terms of purpose related functions and object related processes, whereas, system mangers at a gold level are more likely to be considering system performance at a functional purpose, and values and priority measures level

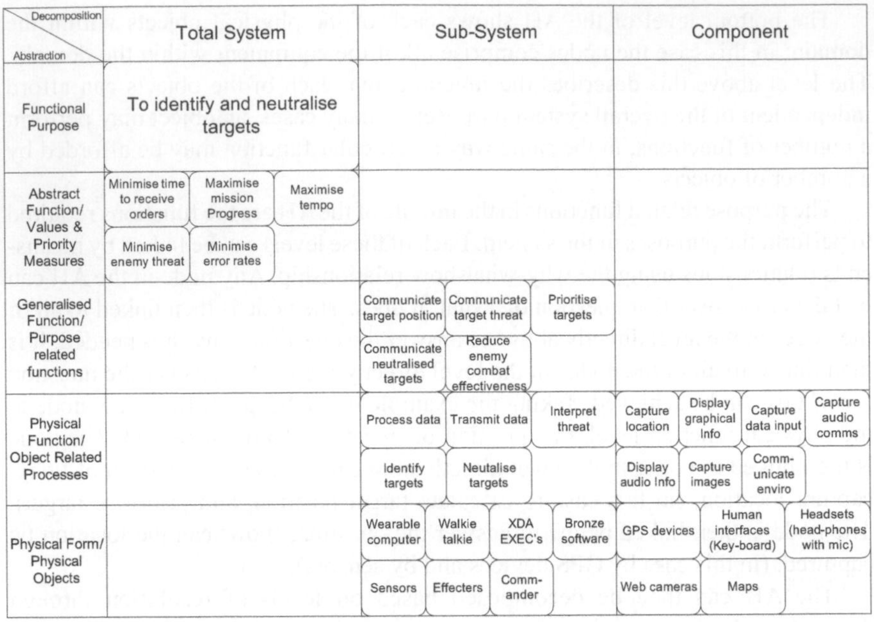

Decomposition / Abstraction	Total System	Sub-System	Component
Functional Purpose	To identify and neutralise targets		
Abstract Function/ Values & Priority Measures	Minimise time to receive orders / Maximise mission Progress / Maximise tempo / Minimise enemy threat / Minimise error rates		
Generalised Function/ Purpose related functions		Communicate target position / Communicate target threat / Prioritise targets / Communicate neutralised targets / Reduce enemy combat effectiveness	
Physical Function/ Object Related Processes		Process data / Transmit data / Interpret threat / Identify targets / Neutralise targets	Capture location / Display graphical Info / Capture data input / Capture audio comms / Display audio Info / Capture images / Comm-unicate enviro
Physical Form/ Physical Objects		Wearable computer / Walkie talkie / XDA EXEC's / Bronze software / Sensors / Effecters / Comm-ander	GPS devices / Human interfaces (Key-board) / Headsets (head-phones with mic) / Web cameras / Maps

Figure 6.5   Abstraction decomposition space for sensor-effecter activity

Control Task Analysis

The Work Domain Analysis phase looked at the domain independent of activity. In order to understand the domain further it is advantageous to look at the known recurring activities that occur within this domain. The second phase of the analysis, Control Task Analysis (ConTA), models one or more of these known recurring tasks, focussing on what has to be achieved, independent of how the task is conducted or who undertakes it.

Control task analysis uses Rasmussen's (1986) Decision-ladder; the ladder in Figure 6.6 can be seen to contain two different types of node, the rectangular boxes represent data-processing activities and the circles represent states of knowledge resulting from data processing. The decision-ladder shows a linear sequence of information processing steps that is folded over. Novice task performers are expected to follow the decision-ladder in a linear fashion (starting in the bottom left corner and proceeding up and over to the bottom right), whereas, expert users are expected to by-pass sections of the ladder based on their previous experience and understanding of the system (see links between legs in Figure 6.6). Individual tasks can be modelled onto the decision-ladder. Figure 6.6 illustrates the task of identifying a target and attending to it. The system is activated when a target is spotted (shown on the bottom left leg). Once a target is spotted, information is recorded on its location and type (no inference or calculation is made). Assessment is then made to calculate the threat of the target. Once a threat has been assigned, the target is then considered relative to the task and the environment and a priority is placed upon it. This prioritisation allows the target to be assigned to an effecter. A target is then identified and finally attended to.

In order to expedite this process it is possible to bypass some of the steps. Removing some of the decision-making processes allows the transition from spotting the target to attending to it to be expedited. Figure 6.6 shows each of the possible leaps (circle to circle) and shunts (circle to square). Figure 6.6 illustrates that the shortest path for this paradigm is that the target is spotted, information recorded and this information is used to attend to the target. The shortcuts are explained in Table 6.1.

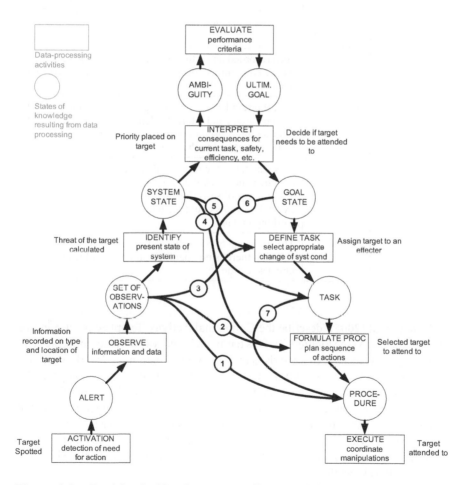

Figure 6.6   Decision-ladder for sensor-effecter activity shortened

Table 6.1   Description of the shortcuts in Figure 6.6

Shortcut	Type	Description
1	Leap	This is the simplest path through the decision-ladder; a target is spotted, information is recorded on the type and location of the target this is then sent directly to the effecter who stops what he is doing and proceeds directly to the target.
2	Shunt	This shortcut is very similar to shortcut 1. The difference is that an additional step is included to allow a procedure for selecting the target. Shortcut 1 assumes that this process is not necessary.
3	Shunt	This shortcut leaves the left leg after information on type and location has been recorded, next an effecter is selected to determine the task to complete.
4	Shunt	This shortcut leaves the left leg after information has been recorded and a threat calculated. This threat is then used to inform the decision about which target to attend to before the execution phase.
5	Shunt	This shortcut leaves the left leg after information has been recorded and a threat calculated. The thereat information is used to inform which effecter should be used to attend to the target.
6	Leap	This leap takes the process after information has been recorded and a threat and priority has been assigned. The process of selecting an effecter is bypassed; examples of this could include sensors and effecters working in pairs.
7	Leap	The final leap bypasses the select target phase, in this situation there is no cognitive process involved in target selection.

Naikar et al (2005), describe the Contextual Activity Template for use in this phase of the CWA (see Figure 6.7). This template is one way of representing activity in work systems that are characterised by both work situations and work functions. According to Naikar et al (2005), the work situations (situations decomposed by schedules or location) are shown along the horizontal axis, and the work functions (activities characterised by its content independent of its temporal or spatial characteristics (Rasmussen et al, 1994) are shown along the vertical axis of the Contextual Activity Template. The dashed boxes indicate in which situations the work functions can occur, whereas, the circles and whiskers indicate where the work functions typically occur.

The work functions captured in this diagram are typically similar to the purpose related functions in the WDA (see Figure 6.4). For the command and control micro-world three distinctly different situations have been selected due to the constraints enforced by their geographical variation, these are; in the field searching for targets; in the command centre; and in the field attending to targets. Figure 6.7 shows that the constraints imposed on the system mean that two of the functions are bound by the situations (records information on the type and location; and attend to the target). The functions of prioritising the targets and of assigning the threat can take place in any situation. Figure 6.7 illustrates that the function of 'calculating the threat of the targets' can occur in any situation; however, it typically occurs in the field whilst searching for targets. The work function 'record information on the type and location of the target' shown in Figure 6.7 is constrained to only being able to occur in the field whilst searching for targets.

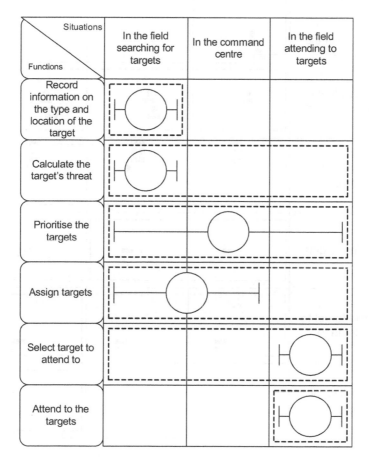

Figure 6.7   Contextual Activity Template

The decision-ladder introduced in Figure 6.6 can be used to communicate which stage of the task is being completed at any particular combination of work situation and function. Figure 6.8 shows the Contextual Activity Template overlaid with this information, it should be noted that these decision-ladders are only intended to be indicative of the typical area of the decision-ladder involved. It is important to point out that this could be completed in a number of ways.

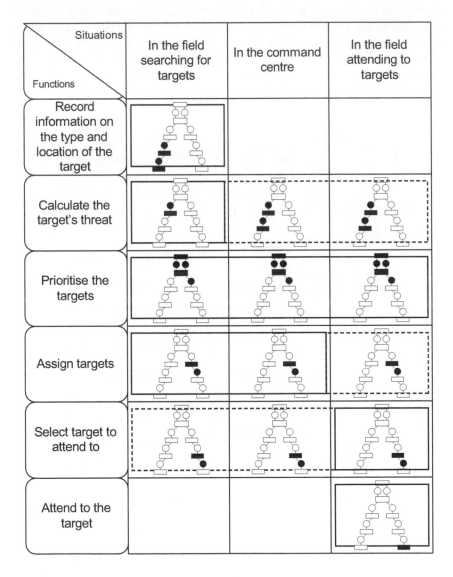

Figure 6.8   Contextual Activity Template with decision-ladders mapped on

The application of the ConTA leads the analyst to consider, for the first time, known reoccurring activity within the domain. Here further system constraints are discovered by considering these typical activities against specific situations. In this example, the acknowledgement that certain activities are bound by geographical location is fundamental to allocation of work.

Strategies Analysis

Strategies analysis is used to look in more detail at known recurring activities. This step of the analysis considers the tasks analysed in the ConTA phase and considers the strategies that may be used to complete them. The strategy adopted by an actor at a particular time may vary significantly. Different actors may perform tasks in different ways; with the same actor also performing the same task in a variety of different ways. There are a number of strategies for achieving the same ends with the system described in the Abstraction Hierarchy. Each of these strategies uses different resources and distributes the workload in different ways. Figure 6.9 shows six of the most common methods for attending to a target. The list is not intended to be exhaustive; it is intended to capture those methods that are common and most likely.

The analysis plots a strategy moving from a start state of an identified target, to an end state of the target being attended to. Use of the decision-ladder representation in Figure 6.9 illustrates a typical path through the process because of selecting a certain strategy. The first strategy shows that the task is completed

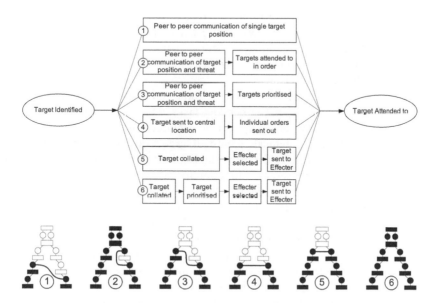

Figure 6.9   Strategies analysis for sensor-effecter activity

at a simplistic level by peer-to-peer communication and without threat calculation or prioritisation. This situation requires the targets to be attended to as they are detected. An example of a more complex situation is situation 6, here the target is processed centrally and considered with all other targets, a priority is assigned and the appropriate effecter selected.

The Strategies Analysis phase of CWA leads the analyst to introduce specific strategies for the first time, based upon the information gathered from the previous phases it is possible to populate these representations quickly.

Social Organisation and Cooperation Analysis

Social Organisation and Cooperation Analysis (SOCA) models the constraints governing the division of tasks between the resources and addresses how the team communicates and cooperates. The objective is to determine how the social and technical factors in a system can work together in a way that enhances the performance of the system as a whole. In order to show who has the capability of doing what, it is possible to map each of the identified actor types (sensor, effecter and commander) on to the existing tools (ADS, Decision-ladder and Strategies Analysis). These can be coded using arbitrary shading (dark, medium and light grey for the sensor, effecter, and commander respectively) to show where each of the actor groups can conduct tasks. The application of colour coding results in a concise graphical representation; verbose annotation is often required to capture the reasoning behind the coding as well as to capture the links between the phases.

Figure 6.10 shows the abstraction-decomposition space (ADS) shaded to show the nodes that can be used by the key actor groups. The total system requirements have been left blank as these are generic and apply to all actors. The diagram clearly shows which of the nodes are specific to individual actors and which of the nodes can be attributed to any actor.

Figure 6.11 shows the decision-ladder introduced in Figure 6.6 shaded to show where each of the actor types can conduct tasks. Due to the limitations of the system, sensors are the only actors that can detect targets, and effecters are the only actors who can attend to previously identified targets (highlighted in Figure 6.7). This leads to the 'feet' of the ladder being shaded dark for the sensors and 'medium grey' for the effecters. In these cases, they are the only actors physically capable of conducting these tasks. The remaining part of the decision-ladder involves taking the basic information from the sensor, interpreting it and making a decision about which targets to attend to. In this case, this activity can be conducted by the sensor, the commander or the effecter. For this reason, the nodes are tri-shaded.

Figure 6.12 shows that the Strategies Analysis diagram introduced in Figure 6.9 can also be shaded to show the actors engaging in the task. Here the initial state must start with the sensor and end with the effecter; however, the strategy used in the middle can be enacted by the sensor, the commander or the effecter.

The SOCA phase captures the constraints enforced by the actor type, each of the three representations from the previous sections illustrates these constraints in a different way. The benefit of this phase lies in the ability to capture an understanding of the constraints surrounding task allocation and the allocation of resources.

Worker Competencies Analysis

The final phase of the CWA framework, Worker Competencies Analysis (WCA), involves identifying the competencies that actors require for performing the required activity within the system under analysis. WCA is concerned with making the task easier for the end user by use of techniques such as mental models.

The Worker competencies analysis phase of CWA can be described by Skills, Rules and Knowledge (SRK) based behaviour. According to Rasmussen et al (1994), Skill-Based Behaviour (SBB) is performed without conscious attention. SBB typically consists of anticipated actions and involves direct coupling with the environment. Rule-Based Behaviour (RBB) is based on a set of stored rules that can be learned from experience or from protocol. During RBB individual goals are

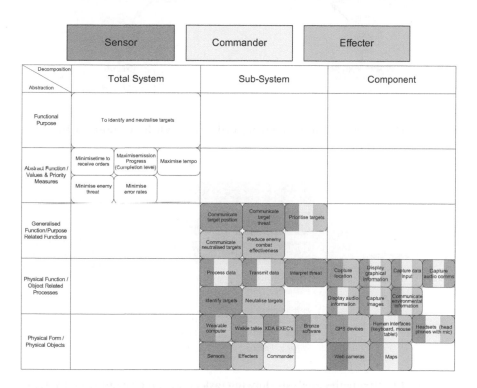

Figure 6.10   ADS showing nodes used by each of the key actor groups

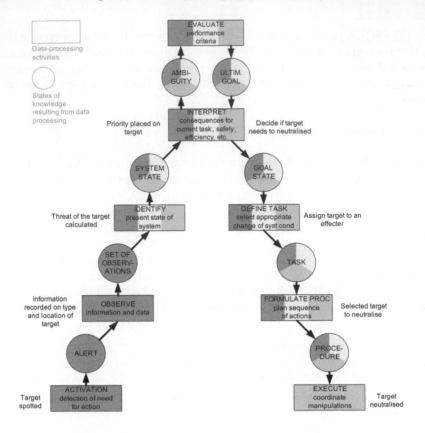

Figure 6.11   Decision-ladder showing tasks that can be conducted by actor types

Figure 6.12   Strategies analysis showing tasks that can be conducted by actor types

not considered, the user is merely reacting to an anticipated event using familiar perceptual cues, unlike SBB, users can verbalise their thoughts, as the process is conscious. When decisions are made that explicitly consider the purpose or goal of the system, the behaviour can be considered Knowledge-Based Behaviour (KBB). KBB is slow, serial and effortful because it requires conscious, focal attention.

The optimum network structure will also be dependent on the behaviour level expected from the actors. The behaviour the actors' exhibit can be classified into three SRK levels dependent on the level of processing required to complete the desired activity. Figure 6.13 shows example responsibilities for each of the stages of the process at the three behavioural levels. The design of an interface should allow actors to perform at any of the three levels; however, by restricting the information shown on a display, the actors can be encouraged to follow protocol rigorously, acting at a rule-based level. More knowledge-based behaviour can be encouraged by providing the actor with additional contextual information promoting the actor to develop a deeper understanding of the current state of the work domain.

Information processing step	Resultant state of knowledge	Skill-Based Behaviour	Rule-Based Behaviour	Knowledge-Based Behaviour
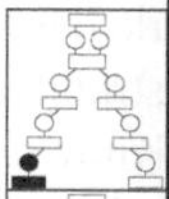 Searching for possible targets	Whether targets are in vicinity	Monitor vicinity for explicit target sightings	Anticipate position based on visual cues from the environment	Infer likely positions
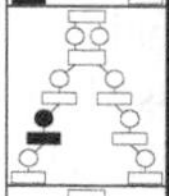 Record information on type and location of target	Understanding of capabilities and location	Direct observations made on target	Experience used to infer capabilities from visual cues	
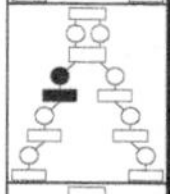 Calculate the threat of the target	Understanding of the implications of the targets capabilities and location	Simple conversion of capabilities to threat	Experience used to infer threat from targets capabilities and location	Target threat considered against overall objectives
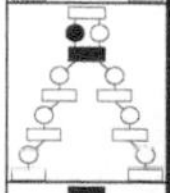 Calculate the priority of the target	Understanding of the relative priority of the targets	Priority based on a single factor such as distance threat	Simple balance applied using experience to decide priority order of targets	Targets prioritised considering the overall objectives and implications
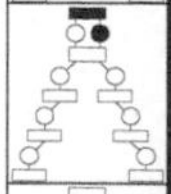 Evaluate implications of neutralising target	Understanding of effects of neutralising target	Consider implications based on correct understanding of situation	Consider implications based on previous experience	Consider implications by hypothesising possible implications
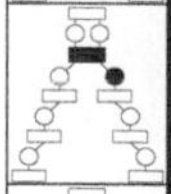 Determine if target needs to be neutralised	Whether target is to be neutralised	Protocol used to decide if target needs to be neutralised	Protocol used along with exception statements to decide if target is to be neutralised	Deviations form protocol considered against overall objective
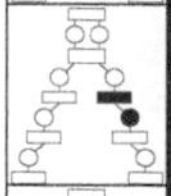 Assign Effecter to target	Effecter assigned to target	Assign target based on single measure such as location workload	Use simple rules to balance workload and location	Assignation based on greatest effect on overall system purpose
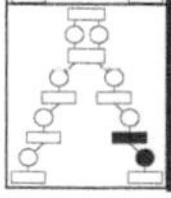 Determine target to be neutralised	Target selected	Targets selected in priority order	Targets selected in priority order unless new information is received	Target selected based on greatest effect on overall system purpose

Figure 6.13   SRK levels for each of the actors (representation adapted from Kilgore and St-Cyr, 2006)

Conclusions

This case study has described the sensor to effecter system at each of the five CWA phases. The analysis has described the domain and answered questions on why the system exists, what it should do, how it should do it and who should be enacting the various stages of the task. The first phase WDA identified the purpose of the system along with the metrics to assess its performance. The analysis also captured the constraints governing how the system could be utilised in the future when faced with unknown unanticipated events. The second phase, ConTA, captured the standard recurring task of detecting and attending to targets as well as the constraints enforced by the situation within the domain. In the third phase the multitude of strategies for completing this recurring task were examined. The fourth phases introduced assigning the stages of the task to actor groups; here a constraint-based approach was used capturing all of the possible organisational combinations to complete the task. Finally, the fifth phase WCA identified examples of the behaviour exerted at each of the three levels identified by Rasmussen et al (1994).

Figure 6.14 shows how the phases of CWA are interconnected in this case; this high level of interconnectivity is one of the strengths of the CWA framework. The domain is first analysed independent of activity in the WDA, here the constraints bound by the functions the physical objects can perform is captured. Known reoccurring tasks can be extracted from the products of the WDA and analysed in detail in the ConTA, the ConTA considers how the constraints imposed by the geographical location of the activity affects what functions are possible. The activities identified in ConTA are explored in detail in the Strategies Analysis

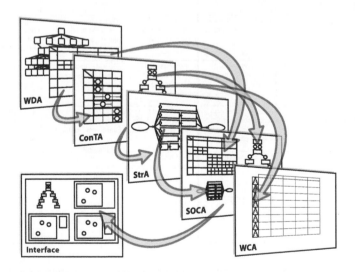

Figure 6.14   Diagram showing major links between phases

(StrA). All three of these phases then feed into the SOCA phase where they are coded to indicate which of the actors can be associated with parts of the process.

The CWA, by focusing on constraints, demonstrates the flexibility of the network: Due to the physical nature of the sensors, they are essential to the system as they are the only method for capturing target positions. The effecters are also essential to the system as they are the only means of attending to targets. The commander/command team has no unique role and is therefore non-essential to the system. The physical actions of 'sensing' and 'attending to' are fixed. However, the more complicated tasks of interpreting, evaluating and defining the task can be assigned to anyone within the system, although as the Contextual Activity Template in Figure 6.7 shows, the situation within the domain is likely to have an effect on the allocation of task.

By changing the roles and responsibilities of the groups of actors, it is possible to compensate for environmental changes by rapidly reconfiguring the network. By focusing on the constraints, the analysis captures every physically possible network configuration. The potential benefits of rapidly reconfigurable interfaces either automated or manual are great in terms of both system efficiency and speed of task completion.

Design for the Sensor to Effecter System

The previous section has analysed the domain, capturing a variety of system constraint based descriptions. In this section, the aim is to use this information to develop bespoke interfaces. By considering the constraints relating to the actor groups the bespoke displays will be developed. The Contextual Activity Template in Figure 6.7 shows that the domain situation has an effect on the type of activity that is conducted, as certain actors are constrained to different situations they are in turn constrained to certain activities, it is therefore applicable to consider developing bespoke interfaces for them. Actors in the domain navigating around the environment (such as the sensors and effecters) are more likely to require features of a physical interface representing lower levels of abstraction, whereas, actors involved with evaluating and manipulating system performance are more likely to require features of a functional interface at higher levels of system abstraction.

The Physical Interface

The physical interface is designed with a strong consideration of the physical objects within the domain; at this level, the emphasis frequently focuses on representing and monitoring the components and their affordances within the physical environment in which the task takes place. Due to its physical bias, much of the design is informed by the two bottom levels of the AH (physical objects and object related processes).

Devices

Each of the actors within the command and control micro-world are equipped with either a computer or a PDA. Unsurprisingly the design of the interface for these displays has a significant effect on the efficiency and the efficacy of the actors within the system. The following interfaces have been developed to be used on a variety of digital devices; a PDA (see Figure 6.15) or a laptop (see Figure 6.16)

Figure 6.15   PDA with stylus

Figure 6.16   Standard laptop

The Starting Point

The decisions ladder from the SOCA phase (see Figure 6.11) led the development of the interfaces by capturing the information requirements for the actors, indicating the contextual constraints associated with certain parts of the decision making process, thus the required information can be identified for each of actor groups for any given network configuration. The decision-ladder breaks down the activity into a number of processes that can be represented in the interface (see Figure 6.6); the interfaces are constructed to show the minimum information on their base display, this can then be supplemented with additional information, as it is needed dependant on the activity required from the actor. Table 6.2 shows that the level of complexity of the display is related to the task being completed with the tasks of identifying and neutralising requiring minimal information and the more considered tasks of selecting prioritising targets and selecting effecters requiring richer information.

Modular Design

A modular design has been adopted to enable interfaces to be changed along with the network. The design of the individual displays is informed by the requirements placed upon the actor; only the information that is pertinent to the activity is displayed. As the roles change, information is added and removed; cues to inform actors that roles and responsibilities had changed were also required from the interface. When the network is reconfigured the unit's roles and responsibilities are also likely to change. An effective interface design should represent this change in network configuration and change in unit responsibility. Typically, when a unit is required to step up to more effortful behaviour their information requirements will become greater. The modular design increases the salience of new information as a new part of the screen becomes populated.

Key functions were extracted from the decision-ladder (see Figure 6.6) and modelled in the Contextual Activity Template (see Figure 6.7). These functions each have specific information requirements that can be captured in an interface. As Table 6.2 shows, there are six functions that can be completed, independent of the actor conducting the activity. These are detect and communicate the location of the target; calculate the target's threat; prioritise the target; allocate the target to an effecter; select the target; and neutralise the target.

Detect and Communicate the Location of the Target

In order to communicate the target position information has to be sent on the location, the interface should facilitate a method for easily recording and communicating the location of a target in respect to the sensor.

Table 6.2   Required display information by activity

	Activity	Required information	Information required on the display
	Detect and communicate Target	Own location (white) and location of the target (grey)	
	Calculate Threat	Location of the target (grey) along with information on the behaviour and weapons capabilities	
	Calculate Priority	Location and threat of other targets (grey) also the location of available effecters (white)	
	Select Effecter	Location of effecters (white) and understanding of their workload	
	Select Target	Location of assigned targets (grey)	
	Neutralise Target	Own location (white) and location of the target (grey)	

One method of entering this information is by clicking on the position of the object (the circle towards the top) in reference to the unit (the circle in the middle) and the environment (buildings and roads shown in the background in grey).

A good interface would support a number of strategies for adding in a target, these would include clicking the location of the target by relating its position in the real world to its position on the map. Another method would include adding in the target by describing its position in terms of distance and heading from current location.

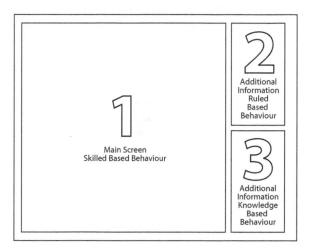

Figure 6.17 Modular interface

Figure 6.18 Location recording interface

Calculate the Target's Threat

To calculate threat information the observation of the target has to be interpreted. The interface therefore requires the same display as Figure 6.18 for the main part of the display. Using a simple set of rules a numerical threat needs to be applied to the object. The interface is required to enable the user to see the threat currently attributed and then increase or decrease it accordingly. In order to support a PDA interface, the ability to communicate a number using the stylus is important, perhaps the simplest way of doing this is via touch sensitive up and down buttons combined with a read out of the current value (see Figure 6.19). To provide the user with flexibility the system should also support other methods of numerical entry such as character recognition and 'virtual keyboards'.

Figure 6.19   Threat calculation interface

Prioritise the Targets

To prioritise the targets, information must be considered against the other targets in the system as well as the overall functional purpose and constraints. A simple interface (essentially the same as the threat interface) can be used to input and represent this priority when it has been calculated.

Allocating the Targets to an Effecter

When allocating the targets, information is required on the workload of the effecters, the interface will need a method of pairing targets and effecters. Figure 6.20 offers an example display where effecters (blue) are linked to target (yellow = allocated, green = neutralised) via a black line. This line can be added using either the stylus or mouse by drawing a connecting line between the two nodes.

Figure 6.20   Threat calculation interface

Selecting the Targets

When selecting the targets, the minimum information required will be the target's positions. Additional information on the threat, priority or locations of other units in the same area may also be of benefit.

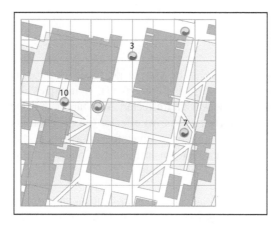

Figure 6.21   Target selection

Neutralise the Target

In order to neutralise the target, the effecter will need to know the targets location a confirmatory message will also need to be sent back to the system when the target has been neutralised.

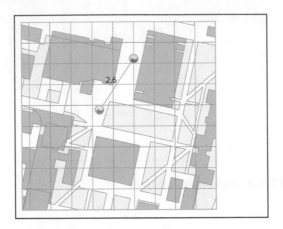

Figure 6.22   Location displaying interface

These displays can be summarised in the Contextual Activity Template (see Figure 6.23) the information requirements for these interfaces have been extracted from the decision-ladder.

Example Networks

Simple peer-to-peer network   The simple peer-to-peer network represents the most basic network possible (see Figure 6.24). It involves direct peer-to-peer communication between sensor and effecter. As a target is detected information is recorded and sent directly to a known effecter (no decision is made on who to send it to). As soon as the effecter receives the information, they move towards the new target. The interfaces shown in Figure 6.25 show the minimal required information for the senor and the effecter.

Advanced peer-to-peer network   In this more advanced network, the sensor also adds in information on the target's threat (see Figure 6.26). The effecter can now see more than one target at any one time; they are required to select the target to move to based on simple rules. When compared with Figure 6.25 the interfaces in Figure 6.27 show the additional requirements placed upon the sensor. The sensor now has to assign a threat to the target. This information is then displayed on the effecter's interface who then considers this threat along with geographical location to decide which target to neutralise.

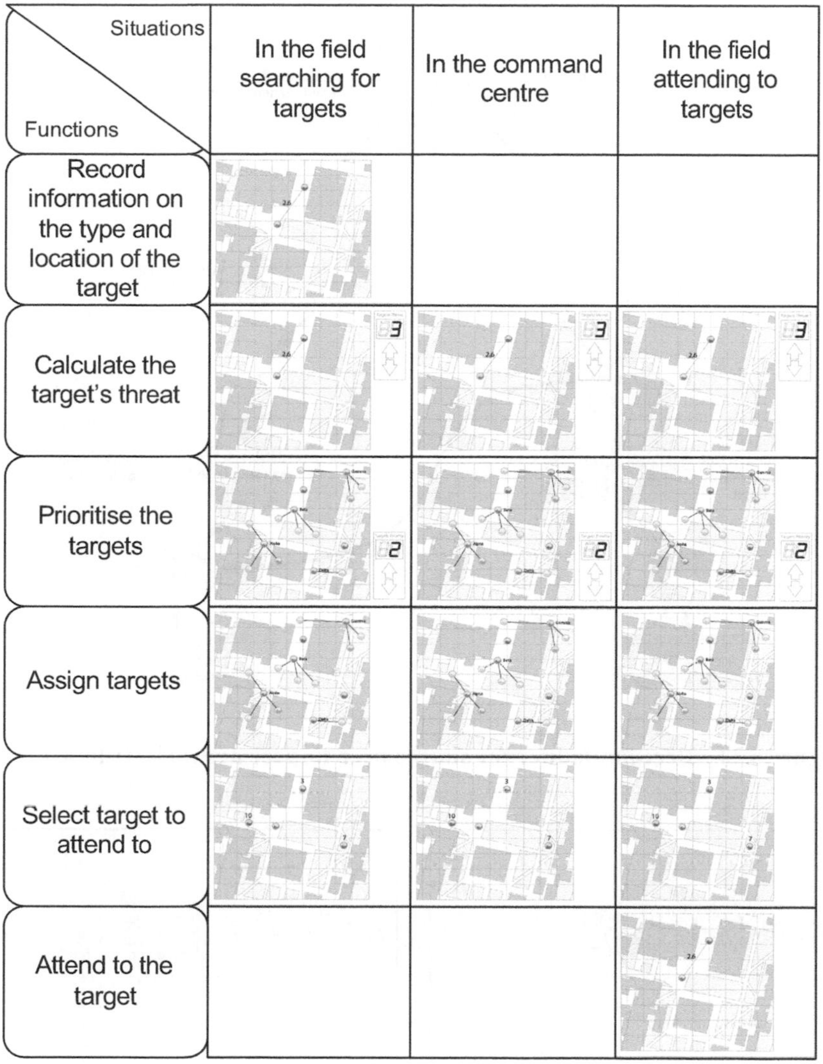

Situations / Functions	In the field searching for targets	In the command centre	In the field attending to targets
Record information on the type and location of the target			
Calculate the target's threat			
Prioritise the targets			
Assign targets			
Select target to attend to			
Attend to the target			

Figure 6.23   Displays shown for each applicable situation and function

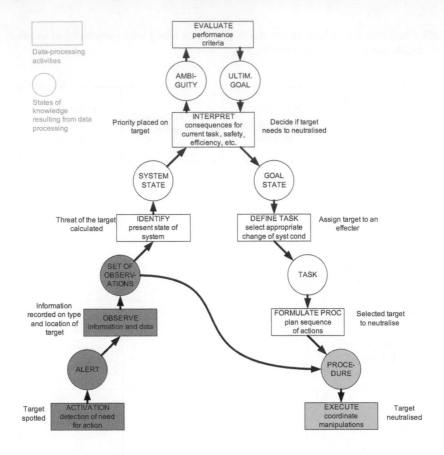

Figure 6.24   Decision-ladder for simple peer-to-peer network structure

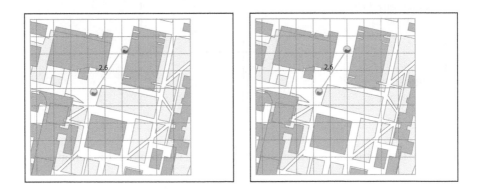

Figure 6.25   Interfaces for simple peer-to-peer network structure (sensor left, effecter right)

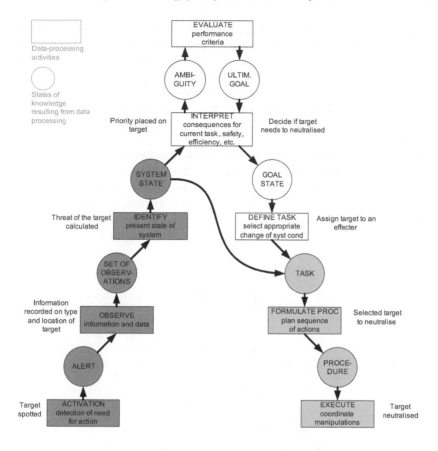

Figure 6.26 Decision-ladder for advanced peer to peer network structure

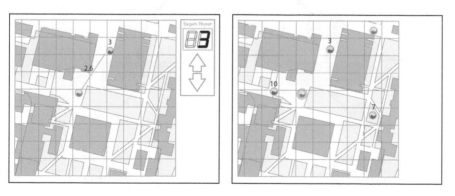

Figure 6.27 Interfaces for advanced peer-to-peer network structure (sensor left, effecter right)

Larger advanced network In a larger network with many sensors and effecters, it is advantageous to have a commanding unit allocating the targets among the effecters (see Figure 6.28); the displays for the sensor and the effecter remain the same as in Figure 6.27 as the same activity is conducted by the agents. Figure 6.29 represents the interface required by the commander for the additional task of assigning targets to effecters. The commander needs to know the location of the targets and the effecters he also needs to have an appreciation of the workload of the effecters.

In order to visualise the effects of system manipulations, a dynamic tool was developed showing the interfaces for each of the three key actors (Figure 6.30); one for the sensors (bottom left), one for the effecters (bottom right), and one for the commander(s) (top right). The displays are manipulated by assigning different parts of the decision-ladder (shown on the top left of the array) to different actors. The decision-ladder can be either assigned to sensor, effecter, commander, or not

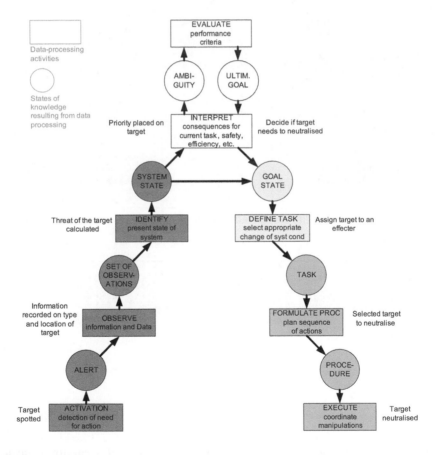

Figure 6.28   Decision-ladder for advanced network structure

assigned (the part of the task is not required). Figure 6.30 shows the simplest configuration. In this situation, the commander is not used in the system, the sensor (shown as the darker ringed circle) has identified a single target (lighter circle) and sent it to the effecter who incidentally happens to be collocated.

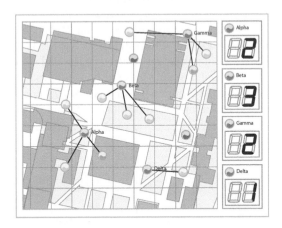

Figure 6.29   Commander's interface for advanced network structure

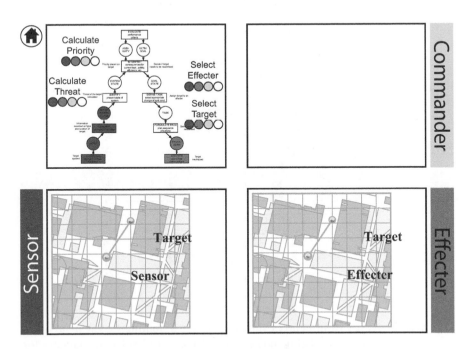

Figure 6.30   Array of interfaces showing simplest configuration

When the sensor is assigned more responsibility, the decision-ladder is shaded to reflect this change in the system. The interface for the sensor is automatically altered in line indicating to the user that their role has changed. In Figure 6.31, the sensor is assigned the additional responsibility for assigning the target's threat and priority. In order to complete this task, the sensor needs to be provided with the location and details of other targets in order to place a relative priority. In Figure 6.32, the sensor is required to allocate targets to effecter; to enable them to do this they need to know the location and workload of the effecters. Figure 6.33 shows the effecter taking responsibility for assigning threat priority, targets and selecting targets. The effecter's display has changed to reflect these responsibilities. The introduction of the commander is modelled in Figure 6.34. In this case, the commander has taken on responsibility for prioritisation and allocation of the targets.

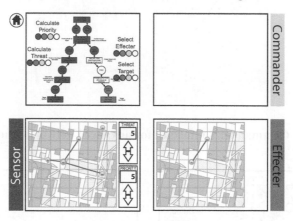

Figure 6.31   Array of interfaces showing sensor assigning threat and priority

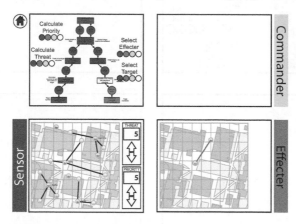

Figure 6.32   Array of interfaces showing sensor assigning threat priority as well as assigning targets to effecters and selecting targets for effecters

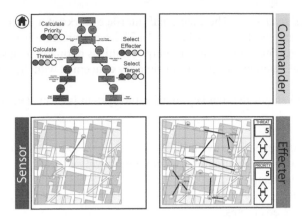

Figure 6.33   Array of interfaces showing the effecter assigning threat priority as well as assigning and selecting targets

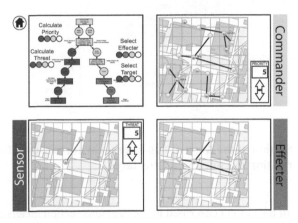

Figure 6.34   Array of interfaces showing the introduction of the commander prioritising and assigning targets

Physical Interface Conclusions

This section has shown how the SOCA phase can be directly used to inform the design of a series of interfaces. By returning to the analysis phase, it is clear that the analysis in the WDA informed the task required and the decision-ladder created in the ConTA phase became the basis for the modelling of allocation of function. The process mapped in this section shows how the interface can be created from this information.

The functional interface   The functional interface works at a much higher level than the physical interface (see Chapter 3). The representation is concerned with how the system is meeting its overall purpose; this is normally judged by a series of metrics. The functional interface is primarily concerned with the top two levels of the AH.

Figure 6.35 shows that the key metrics the system is evaluated against are as follows.

- Minimise time to receive orders
- Maximise mission Progress (Completion level)
- Maximise tempo
- Minimise enemy threat
- Minimise error rates.

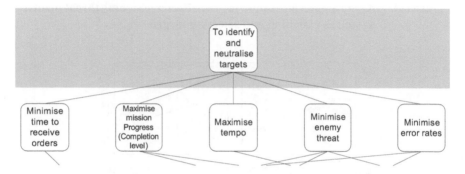

Figure 6.35   Section of the Abstraction Hierarchy for sensor-effecter activity relating to Functional Interfaces

The flexibility of the system described in the previous section means that there are 256 system configurations (there are four parts of the decision-ladder; each can be in four states of: sensor, effecter, commander or none, giving 44 = 256). The system configuration selected will be dependent on a number of factors including:

- Number of units: – larger systems are likely to become more complex, there may be a point where the system starts to perform better with some kind of centralised 'deconfliction'.
- Ratio of effecters to sensors: – the ratio is likely to influence the way the workload is divided between the two types of actor with one group taking over tasks from another to reduce bottlenecks.
- Ratio of targets to sensors: – the number of targets per sensor is likely to be a trigger for a network change in some networks. At a point where the targets become too numerous it may be advantageous for the system to revert to a centralised assignment of targets to sensors.

- Complexity of task: – if there are a number of conflicting requirements, the system may benefit from a higher more formative decision-making process. This decision making process may not be required if the complexity of the task changes, then a quicker rule based system could be adopted.

The role of the gold functional interface to represent the high-level performance measure of the system and to facilitate a method of manipulating the variables described above. It is possible that much of the reconfiguration could be automated based on the development of optimised system formulas. In this case a computer could detect trigger points for system changes and automatically disseminate this change to the actors by reconfiguring their interfaces.

Determining Data Representation Method

The data representation method can be directly informed by the Abstraction Hierarchy; specifically the top two levels (see Figure 6.35). As manipulations are made to the system their implications need to be examined. Perhaps the simplest way of doing this is by tracking the trend of the metrics identified in the values and priority measures level of the Abstraction Hierarchy. The choice of display for each of the metrics identified can be informed based upon the criteria identified for single displays in Chapter 3.

- Is it binary (2 states) / multi-state / analogue?
- Is it between limits?
- Are the limits fixed or variable?
- Is it critical to monitor?
- Does trend need to be tracked?
- Does rate of change need to be tracked?

Minimise Time to Receive Orders

The data for 'minimise time to receive orders' is analogue; it can take any value from zero to infinity. It is likely that there will also be a point at which the time to receive orders will be considered unacceptable. The data can be considered as between limits; however, the limits (the upper limit) are not fixed.

Figure 6.36   Data limits for time to receive orders

The actual numerical time taken to receive the orders is likely not to be as important as the context it is placed in, has it increased or decreased (its trend and rate of change). By adding a history, it is possible to examine the effect of changing variables within the system. It may also be important to compare the current value against a target maximum value.

Maximise Mission Progress (Completion Level)

The data for 'maximise mission progress' is analogue between limits (no progression to completed) and the limits are fixed.

The difficulty in this particular example is that the mission completion level cannot be based on targets eliminated, the precise number of targets in the field will not be known until the end of the exercise. In order to quantify mission completion, data would need to be recorded on another measurable such as percentage of the battle-space (in terms of geographic area) swept.

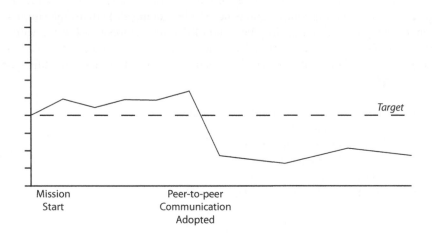

Figure 6.37   Example of a trend chart showing time to receive orders plotted on same axis as target level

Figure 6.38   Data limits for mission progress

Maximise Tempo

Tempo is normally defined by the number of actions per unit of time, for example in music beats per minute. In this case, tempo could be measured by targets neutralised per hour. Essentially the tempo is measuring the rate that jobs are completed. To gain a full picture, it is also beneficial to monitor the tempo of targets sensed, so that an understanding of system balance can be gained.

Figure 6.39 shows an instantaneous view of the situation. In order to show trend, a stack graph can be used, an example of this is shown in Figure 6.40.

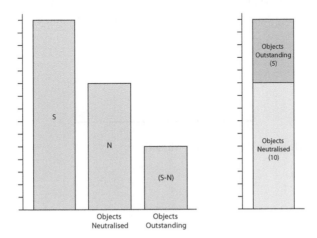

Figure 6.39   Example of data representation for 'tempo'

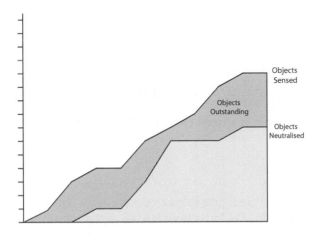

Figure 6.40   Stack chart showing objects sensed and objects neutralised, the dark section between the lines shows the outstanding targets

Minimise Enemy Threat

The data for 'minimise the enemy threat' is very similar to that of the time to receive orders. The data lies between limits; the lower limit is fixed at zero; however, the upper limit is not fixed.

The level of enemy threat at any one time can be calculated by summing the threat of all of the identified targets in the domain at any one time. By neutralising the targets in an indiscriminate order the threat level will be reduced; however, by targeting high value targets the overall threat level can be more rapidly reduced.

It may also be beneficial to show the average threat of a unit in the field so that the relative threat level can be compared as new targets are added and neutralised.

Figure 6.41   Data limits for enemy threat

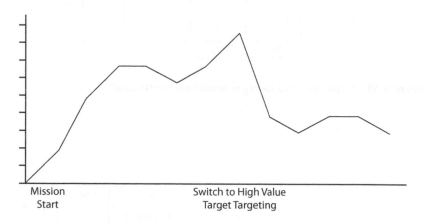

Figure 6.42   Example of a trend chart for domain threat tracking

Minimise Error Rates

As with 'enemy threat' and 'time to receive orders', the data for 'minimise error rate' lies between limits. The lower limit is fixed at zero; however, the upper limit is not fixed.

Any errors occurring need to be considered in context with other events. The detection of an error is also far more important than a cumulative value therefore a mark on the graph may be more beneficial than tracking the number using trend lines. It may be possible to detect common themes happening just before an error takes place using this representation.

Synthesis of Displays

From the means-end links in the Abstraction Hierarchy, it is clear that all of the metrics discussed have the potential to influence the functional purpose. As introduced in the start to this chapter it is also apparent that some of these metrics may be contradictory; an improvement in one may cause a degradation of another. It is therefore advantageous for the metrics to be shown on the same workspace. The type of display informs the layout of the metrics, it is sensible to group all the displays compared against time into a cluster stacked vertically. The maximise tempo and maximise mission progression form a further two independent information groups.

Figure 6.45 shows the first attempt at grouping the displays. In the top left hand corner of the display, the mission progression is shown as a meter. The main part of the screen is taken up by the trend charts cluster made up of two trend charts, the top one shows the outstanding targets in blue and the domain threat level in green each having their own respectively coloured axis. It can be seen that there is a correlation between the two; however, it is not directly related as different

Figure 6.43   Data limits for error rates

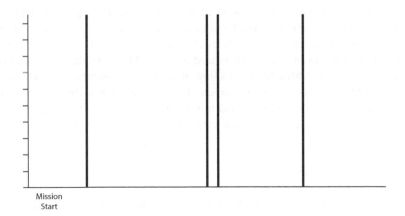

Figure 6.44   Chart showing the tracking of individual error occurrences

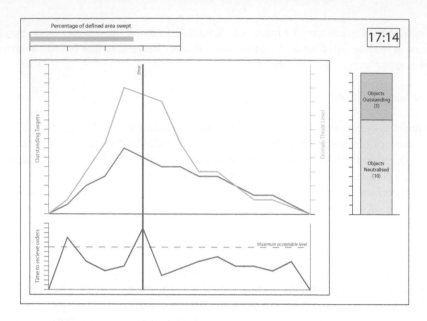

Figure 6.45   Synthesis of the display

targets have different threat levels. The trend chart below shows the time to receive orders; this has been plotted in line with the chat above so that the effects of spikes in either graph can be compared at the same instance. Superimposed over both trend graphs is a red line to signify the time at which an error occurred. On the right-hand side of the display, current instantaneous information is given on the outstanding and neutralised targets. A clock is also provided in the top corner of the screen.

Figure 6.46 shows a 2nd iteration of the display. A stack graph is added so that the trend of the ratio of sensed and neutralised targets can be examined. The mission progression remains outside of this cluster; however, to make effective use of space, it has been rotated and placed on the right-hand side. Digital readout displays have been added to the display. Instantaneous values for each of the monitored parameters remove the burden of interpreting the scales from the user. An additional section has been added to record key events on the time line. The display has also been scaled so that it would fit on a monitor without the need for scrolling.

Controlling the Display

The display shown in Figure 6.46 allows the metrics influencing the functional purpose to be monitored; however, the interface does not contain a control element allowing the person monitoring the display to make any changes. The display

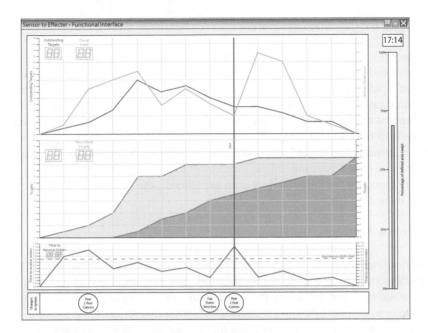

Figure 6.46   2nd iteration of display synthesis

in Figure 6.47 includes controls to manipulate the variables within the system. The buttons are colour coded and located close to the displays that they directly influence. Controls are supplied to add or remove sensors (yellow arrows) and add or remove effecters (blue arrows). Buttons are also included to switch between peer-to-peer communication and hierarchical communication. All of the controls available to the user are located in one group.

Changing the Number of Units

As identified in the SOCA phase the allocation of resources and activity affects the way in which the activity is conducted. Increasing the number of units decreases the time taken to complete the task as more units can be searched and neutralised in a given period. By changing the sensors and effecters independently, the ratio can be changed; this allows the system to be optimally balanced. It is inefficient to have units in the field that are not working at full capacity. There will be a significant lag between activating the controls and seeing the results on the display. This is due to the time taken for the sensors to enter the domain and find targets and for the effecters to reach and neutralise a target.

Changing the Command Structure

As shown in the Strategies Analysis, the command structure affects the strategy adopted and the path taken through the decision-ladder. The command structure can be rapidly changed between hierarchical and peer-to-peer. This change can be communicated to the units through changes to their respective physical interfaces. Dependant on the number of units in the field, the choice of command structure is likely to have a significant effect on time to receive orders and the threat level within domain. The threat level is expected to be lower in a hierarchical system as there is a central command able to direct units to high value targets.

By changing the variable during an experiment, the system can be optimised to be more efficient.

Functional Interface Conclusions

The functional interface is formed at a higher level of abstraction that the physical interfaces used by the units within the field. The domain is considered in terms of the metrics that influence the overall functional purpose. This section has shown how an interface can be developed from a theoretical underpinning. The Values and Priority Measures become the metrics for system performance; these are tracked in the proposed interface on a number of trend charts. The design of these trend charts are informed by a number of design heuristics based upon their limits and required context these can then be clustered together based upon further design heuristics.

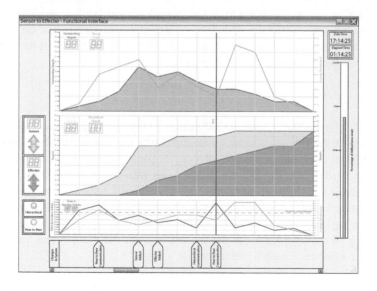

Figure 6.47   Display with controls added

Chapter Summary

This chapter has introduced a complex sociotechnical system and applied CWA to its analysis; the analysis has covered all five phases of the framework; highlighting the benefits of extending the analysis. The second half of the chapter explains in detail how the products of the analysis have been used to inform a series of interface designs. The analysis directly informed the design and development of a number of interfaces, tailored to each of the actors; the situation specific information requirements for each display were elicited from the tasks defined in the decision-ladder. This methodical, analytic approach to the design and development of the displays creates a clear 'audit trail' of the design decisions made. This in part removes some of the 'black art' traditionally surrounding interface design.

By focusing more heavily on the fourth phase (SOCA), this chapter provides further validation for CWA as a comprehensive framework of tools. As described earlier, the majority of previous CWA applications have utilised only the WDA phase. It is contended that future analysis should utilise the phase that is most applicable to both the research aims and the domain in question. Although the SOCA phase has led the design in this case, this phase would not have been completed without the information captured in the preceding phases. This analysis therefore demonstrates the interconnection between the CWA phases and reinforces the benefits of extending a CWA beyond the initial Work Domain Analysis and Control Task Analysis. Further, this chapter provides evidence of how CWA, when applied in its entirety as a suite of tools, can be effective in informing the design of systems and interfaces. The framework is often criticised for not providing sufficient guidance to analysts wishing to design novel systems. One reason for this could be that analysts are either using inappropriate CWA phases or that they are not fully completing the analysis. This analysis demonstrates that, when CWA is undertaken appropriately (i.e. the relevant phases are utilised) and combined with a set of basic design heuristics, clearly defined traceable design requirements can be more easily extracted from CWA outputs.

Chapter 7
Designing Interfaces Using CWA

Chapter Introduction

The aim of the chapter is to investigate how well the CWA framework can support the design and development of a military decisions support tool. The chapter starts by exploring the overlap between CWA and the military planning process, the methods are contrasted and opportunities for synthesis identified. A Battlegroup level example is used to highlight the points raised in this section. A number of conceptual design ideas are then generated from the fusion of these techniques. The developed concept interfaces are then evaluated via a dual-phased experimental process. The first phase of experimentation examines the participants' ability to make decisions based upon previously generated interfaces (static displays presented sequentially). The second phase addresses the participants' ability to make decisions related to a route selection problem, when provided with dynamic reconfigurable interfaces. The experimentation phase, involving 59 participants, collected data on time to make decisions, certainty of decision, and ease of use to evaluate the displays. The results show that each of the displays performed well, and were rapidly interpreted by the users without the need for formal training. The results also identified statistically significant differences between the derived displays; therefore, making it appropriate to draw wider design recommendations.

The wider question of the applicability of digital support tools to support a cognitive process, such as decision-making, is not addressed in this chapter. There is a wide body of research devoted to it; warranting the need for international conferences. Theories, such as Naturalistic Decision Making (Zsambok and Klein, 1997), bring into question the entire concept of CoA selection as a decision support element. Naturalistic Decision Making studies, involving expert decision makers playing chess (Klein, et al, 2005), found that expert decision makers generated the optimum decision as their first option, thus negating the need, and benefit, of generating multiple options. These models, however, more often than not, are focused on individuals rather than the decision-making teams. Whilst it is acknowledged that CoA selection may not be required in all decision-making activities, it is perceived that, in many cases, there may be benefit in developing a number of CoAs, exploring their viability, and comparing their suitability.

New Warfare

Warfare is changing; current operations in Iraq and Afghanistan could scarcely look more different from images of 'symmetric' warfare; gone are the days of defined battlefields of relatively equal forces engaging each other. Modern soldiers are now required to provide humanitarian assistance, peacekeeping, as well as engage in traditional warfighting. More so now than ever, military decision makers are required to consider, in more detail, the implications of their action; not only in terms of military, but also in terms of diplomatic and economic consequences. In such domains, the relationship between action and reaction is notoriously difficult to predict, decision makers are required to consider an ever-increasing number of factors. A number of approaches have been developed by defence organisations around the world to address this complexity. The aim of this chapter is to explore how these approaches can be combined with the CWA approach to develop further insights into the development of digital decision support tools.

Effects Based Operations (EBO)

The effects based approach has recently received a lot of attention within military and political circles. The approach advocates a move away from a sole focus on militaristic issues, to a consideration of the diplomatic, information, militaristic and economic effects of a decision at all levels of the command chain. Arguably, military commanders have been using intuitive versions of the effects based approach for hundreds, if not thousands, of years. The following two and half thousand-year-old quote from Sun Tzu's 'The Art of War' captures a form of effects based thinking.

> Those skilled in war subdue the enemy's army without battle. They capture his cities without assaulting them and overthrow his state without protracted operations. (Sun Tzu)

The purpose of the effects based approach is to supplement the subjective, intuitive decision making process with to a more structured rigorous process. With the move from symmetrical (attrition based warfare) to asymmetrical warfare, the focus on effects based operations becomes even more apparent. Smith (2003), a strong advocate of the effects based approach, points out that, 'conventional' 20th century war fighting relies on destruction and, thus, on the existence of some recognised state of hostilities. He uses the terms 'means' and 'will' to highlight the difference between asymmetric and symmetric warfare defining symmetric warfare as situations where there is a relative balance of both means and will. Smith (2003) points out that in a contest between an entity that has both great means and great will, and an entity that lacks one or both, the side with both great will and means is bound to prevail. The outcome is likely to be swift where the challenger's will is weak and his means lacking. It may be less swift, but it will be just as sure, where the means are available but the will lacking, or where the will is strong but the means lacking. Smith (2003) goes on to say that when the contest is between one power that has great means and limited will and another that has limited means but great will, the result is likely to be far from being either certain or swift, with both sides exploiting their individual strengths and exploiting their opponents weaknesses. Classic examples of this include conflict in Vietnam, and the Soviet experience in Afghanistan.

The aim of the effects based approach, therefore, is to transfer the emphasis from the concentration of resources from destroying means to destroying will. According to the EBO Prototyping Team (2005), action is derived from a top down process. Traditional, overarching national and/or coalition goals are established, objectives for military operations are derived from them, and then discrete actions, which are perceived to contribute directly to those objectives and therefore indirectly to the achievement of the overarching goals are planned, executed, and assessed. They go on to comment that the connection between those goals and actions is, too often, drawn intuitively – based as much on the subjective notions as on objective reason.

Essentially, as Smith's (2003) definition shows, EBOs are involved with shaping behaviour:

> Effects-based operations are coordinated sets of actions directed at shaping the behaviour of friends, foes, and neutrals in peace, crisis and war … effects-based operations are not simply a mode of warfare. They encompass the full range of actions that a nation may undertake in order to induce a particular reaction on the part of an opponent, ally, or neutral. They represent a unified approach to national strategy that is as much at the root of peacetime operations as it is of wartime operations. (Smith, 2003, p.47)

The EBO Prototyping Team (2005) show synergies with this view, stating that:

> The effects-based approach features a combination of military and other activities on influencing the overall behaviour of other actors.... In an operational environment, its application allows the planning, execution, and assessment of those operations to be based on a holistic and dynamic understanding of those and others in that environment. The resulting benefits are a set of actions that are explicitly linked to a set of strategic goals (expressed as an 'end state'), coherently harmonised with those of other governmental organizations, and made truly adaptive within the course of their execution by effective assessment.

According to the Joint Doctrine and Concepts Centre (JDCC) (2006) the Effects Based Approach is defined as:

The way of thinking and specific processes that, together, enable the integration and effectiveness of the military contribution within a Comprehensive Approach.

The Comprehensive Approach

It is widely acknowledged that the prevention or resolution of crises can only be achieved by a combination of dealing with the immediate symptoms of crisis and addressing the associated causes within the context of each individual situation ... a multi-disciplinary and multi-agency approach is required to ensure that the most appropriate overall strategy and measures are applied to coordinate and focus activity, in order to offer the best chances of successful resolution. (JDCC, 2005)

The Comprehensive Approach comprises three inter-related elements:

1. A common way of thinking, focused on long-term outcomes, which emphasises the necessity for a systemic, detailed understanding and dynamic assessment of the individual situation at every level and across every dimension.
2. As far as can be achieved, a harmonised, collaborative way of working across government and between nations, International Organisations (IOs) and Non-Government Organisations (NGOs) to enable the production and delivery of integrated strategies to resolve conflicts or crises.
3. The implementation of a comprehensive response, in the interests of international stability and national political objectives, through rigorous situational analysis, planning, execution and assessment.

How do We Judge the Effect on Behaviour?

The ability to judge the effect of a particular action can be very difficult in an effects-based environment. Smith (2003) points out that because the target is human behaviour, the results are not incremental but non-linear. Instead, the process revolves around relentlessly grinding down the enemy until they can take no more.

Behaviour is often shaped by shaping the society in which the intended 'targets' reside. According to the JDCC (2006), the human security agenda requires a response that is sensitive to the extensive, particular needs of societies, communities and individuals. To this end, all constituent parts of a society (rule of law, education, commercial, humanitarian and health, information, military, economic, and diplomacy and governance) should be considered, as well as the history and culture of an individual society. According to Smith (2003) unlike the attrition-based approach, the effects based strategy is conceived and executed as a direct assault on an opponent's will and not a by-product of destroying his capability to wage war. For this reason, the role of the media and information is no longer that of an ancillary support for morale as in attrition-based campaigns, but as a central part of the effort to assault the public will.

Synthesising CWA and EBO

Returning to the EBO Prototyping Team (2005) description of EBO, it is clear that the relationship between overarching national and/or coalition goals, objectives for military operations, and discrete actions can be described by a top down hierarchy. It is possibly to use the Abstraction Hierarchy, used in CWA, to model this relationship; the AH provides a template for exploring the structural means-ends links between the different levels of abstraction. Using the 'why-what-how' triad, any level can be taken to indicate 'what'; the level above will indicate 'why' this is required; and the level below will indicate 'how' it can be achieved. Figure 7.1 illustrates graphically how the EBO terms map on to the CWA Abstraction Hierarchy.

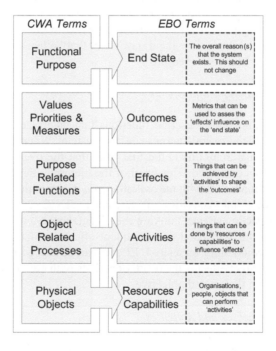

Figure 7.1   Diagram showing how EBO maps onto the CWA Abstraction Hierarchy

Functional Purpose and End States

The 'Functional Purpose' is the reason the system exists, if there is no functional purpose the system is obsolete. The functional purpose should be consistent; it should not change over time (although it may be refined and modified as the system develops). There can be more than one functional purpose. In terms of EBO, the functional purpose can be considered to be synonymous with 'End State'. In true WDA terms the functional purpose should not be a goal; however, when the end state is achieved the system no longer exists in its current form (a new system would be born possibly out of the same components), therefore the end state is a purpose for the duration of the system life.

Values and Priority Measures and Outcomes

'Values and Priority Measures' (VPM's) assess the system's compliance with the functional purpose; the level contains a list of key measures that will have an influence on the functional purpose of the system. In EBO terms, these VPM's can be considered roughly synonymous with 'Outcomes'. Care should always be taken when assigning measures to a system, a version of Goodhart's (the former chief economist of the Bank of England) law states that 'When a measure becomes a target, it ceases to be a good measure.'

Purpose Related Functions and Effects

'Purpose Related Functions' describe functions that take place within the work domain in terms of the overall purpose of the system. When working up from the bottom of the hierarchy this is the first time that the overall purpose of the system is explicitly considered. This level of the hierarchy ties together physical objects and their capabilities with the overall raison d'être of the system. In a very similar way, 'effects' link 'activities' to 'outcomes'. JDCC (2005) points out that in planning and selecting effects, thinking should encompass not only the intended effect(s), but also the associated or subsequent effects, which may be intended or unintended, favourable or undesired. Effects can be categorised as direct and indirect effects. A direct effect is the result of actions with no intervening effect or mechanism between the act and objective.

Object Related Processes and Activities

'Object related processes' are things that can be performed by the 'Physical objects' at this stage they are functions independent of the system purpose and expressed in generic terms. Activities are expressed in a very similar way in EBO with activities being far more generic. It is often the case that one or more 'generic' activities are performed to enable a more specific 'effect'.

Physical Objects and Resources / Capabilities

At the lowest level of the hierarchy lies the 'Physical objects', these are the physical attributes of the system. Or in EBO terms the 'Resources and Capabilities' these may take the form of individuals, pieces of equipment, documents, organisations (both national and international; militaristic and non-militaristic) governments, and charities. At this level, the physical objects are merely described in terms of their structure and form. Capabilities are not considered until the object related processes level.

Benefits of This Type of Representation

The use of the Abstraction Hierarchy has a number of benefits for modelling EBO. The explicit use of a hierarchy encourages the analyst to consider how the different terms relate to one another. The use of the AH as a representational method is rapidly increasing, there are a strict set of rules that accompany the tool, assisting the creation and preventing analysts from misusing the diagram. The use of means-ends links within the AH allow the implications of changes to be considered. Here, either, one-to-many, many-to-one, or many-to-many links can be expressed (see Figure 7.2). Here, outcome 4 and effect 8 are linked by a one-to-one relationship, meaning that if effect 8 were not achieved, it would have a significant impact on outcome 4. Likewise, if outcome 4 were no longer desirable, effect 8 should be cancelled. Outcome 1 shows a more complex relationship with

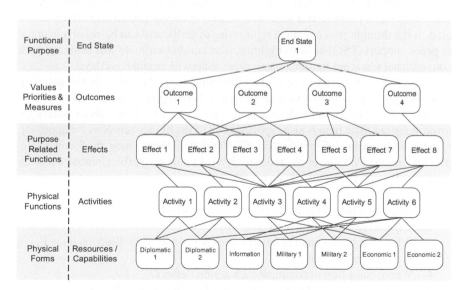

Figure 7.2   Means-ends links

three effects influencing its outcome. If an effect were no longer achievable, further examination would be required to determine the implications for outcome 1.

The domain should be modelled so that the effects are independent of the resources required. In this way, the bottom two levels of the system can be removed and replaced with entirely new resources and capabilities. The overall end state and the outcomes remain the same.

According to JDCC (2005) Those wishing to intervene in a particular situation, or influence events within it, will require extensive visibility and understanding of the features of the operating environment, of which they will themselves become a part. Therefore, an understanding is required of factors, across all dimensions of the environment, which are liable to change and are susceptible to influence. It will be necessary to detect, identify and understand the subtle inter-relationships and interactions that can plausibly exist within a situation, especially with regard to human activity and motivation. It is strongly contented that the Abstraction Hierarchy provides and ideal template for this exploration.

A Contemporary Planning Approach

A case study of a contemporary planning approach is presented to investigate the applicability of the proposed modelling approach. In the British Army battlefield, planning is based around an approach called 'The Combat Estimate'. The Combat Estimate was introduced in 2001 to simplify and speed up the planning process at Battlegroup (BG) level (Group LWCT 2005). The Combat Estimate focuses all of the work strands that are carried out during planning and aims to ensure that all such work has a purpose and leads to a timely, enemy focussed and effects based plan. It is a thought process, not a rigid series of drills, and can be equally adapted to peace support (PSO) as to warfighting. The combat estimate was designed with conventional planning in mind, using paper map with acetate overlays.

The process is meant to offer the Command Staff flexibility in the planning process, allowing them to take a set of orders from higher echelons and use data, intelligence and maps to produce a plan. This in turn leads to sets of orders for lower echelons. The plan is also reported back to the higher echelons for approval. Further background information on the Combat Estimate may be found in the previous report (Walker et al, under review). In summary, the Combat Estimate requires planners to address seven questions:

1. Question One: What is the enemy doing and why?
2. What have I been told to do and why?
3. What effects do I want to have on the enemy and what direction must I give to develop my plan?
4. Where can I best accomplish each action / effect?
5. What resources do I need to accomplish each action / effect?
6. When and where do the actions take place in relation to each other?

7. What control measures do I need to impose?

An example scenario discussed is taken from a training exercise conducted by the author of this book at CAST in Warminster. The description focuses on the key products produced that will inform the construction of the Abstraction Hierarchy.

Once questions one and two have been completed by the planning team, the commander uses the communicated information to construct an intent schematic (see Figure 7.3). The Intent schematic describes in broad terms what should take place. The intent schematic is not drawn on a map; however, it does have geographic relevance. The intent schematic is described by the effects the commander wishes to have on the enemy.

From the intent schematic a number of Courses of Action (CoA) can be developed, at this stage the detail is added to the plan. Here each of the effects is resources with the available units, any time bases conflicts are also resolved. In this example, two CoAs were developed in parallel by different parts of the team on CoA development boards (Figure 7.4), one team adopted a risky CoA moving recourses around the field in order to concentrate fire power, the other CoA, divided the resources, reducing the need for movement and coordination. In the process of developing the CoAs the overall end-state is explicitly considered. During the process, the staff are also considering a number of measures of performance, such as, risk, simplicity, concentration of force, surprise, reserve and C2 less explicitly. In order to evaluate and select the CoAs the options are evaluated against the performance measures. Figure 7.5 shows an example of the

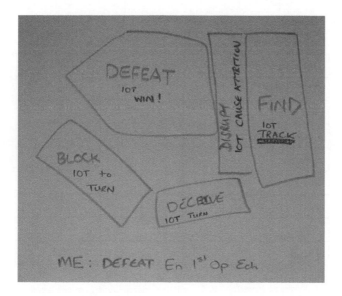

Figure 7.3   Intent schematic

decision matrix currently used. The decision criteria (measures of performance) are listed down the left hand side. This simple rating method gives a numerical value to each of the CoA (ratings of 1 to 3; where 3 is the most desirable). Care has to be taken when using this method, as the individual criteria are not ranked in this matrix; this indicating that each of the criteria is of equal importance. By adding a weighting coefficient, which is multiplied through, it is possible to represent the relative importance of the criteria (see Figure 7.6).

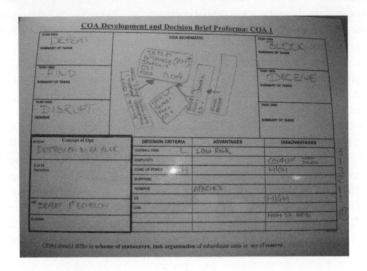

Figure 7.4   CoA 1

Decision Criteria	CoA 1	CoA 2
Overall risk	3	1
Simplicity	1	3
Concentration of force	3	1
Surprise	2	2
Reserve	1	3
C2	1	3
CSS		
Total	**11**	**13**

Figure 7.5   Decision matrix

Decision Criteria	Weighting	CoA 1		CoA 2	
Overall risk	3	3	9	1	3
Simplicity	3	1	3	3	9
Concentration of force	3	3	9	1	3
Surprise	2	2	4	2	4
Reserve	2	1	2	3	6
C2	2	1	2	3	6
Total		29		31	

Figure 7.6   Decision matrix (including weighting factors)

To highlight the differences between the two plans they are plotted on the same board (see Figure 7.7). The commander is then required to select the CoA (or select a synthesis of the two).

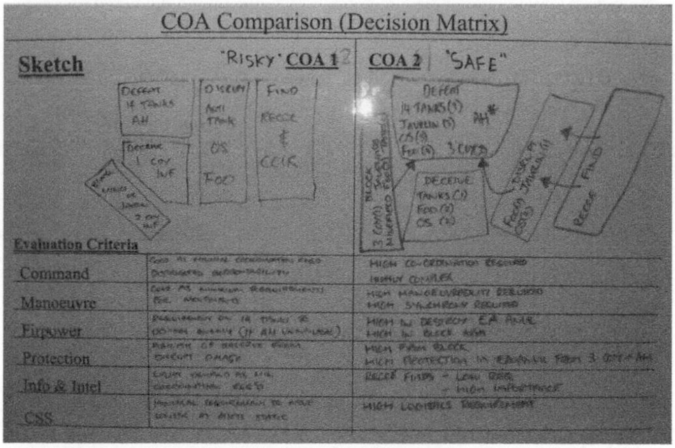

Figure 7.7   CoA comparison (Decision Matrix)

Once the commander has selected the CoA (in this case CoA 2), the activities required are then synchronised in the sync matrix (see Figure 7.8). This process involves determining when each activity needs to take place. The starting point for this calculation is the anticipated time of contact (D-hour); the activities are worked back from this time.

The Abstraction Hierarchy can be used to show the relationship between the overall aim of the system and the individual. Much of the content for the Abstraction Hierarchy can be directly extracted from the products created at CAST (see Figure 7.9). The overall aim of the system is to prevent the enemy passing through. The components at the bottom of the hierarchy can be extracted from the list on the left-hand side of the sync matrix (see Figure 7.8); these include; Tanks, Attack Helicopters, Infantry companies, Anti-tank, Recce units, Engineers, FOO (Forward Observation Officers), Javelin, Artillery, and the Commanding Officer. The physical activities that the units afford can also be found in the main section of the sync matrix (see Figure 7.8); they include engage enemy, detect enemy and mine-laying. These actions can in turn contribute the effects, listed in the commander's effect schematic (see Figure 7.3), such as deceive, neutralise, find, disrupt and block. The effectiveness of the system can be evaluated against a number of Values and Priority Measures similar to those used when evaluating the CoAs, these can be extracted from the 'decision criteria' in Figure 7.4. The result of this Abstraction Hierarchy can be seen in Figure 7.10. Means-ends links can be added to indicate affordance.

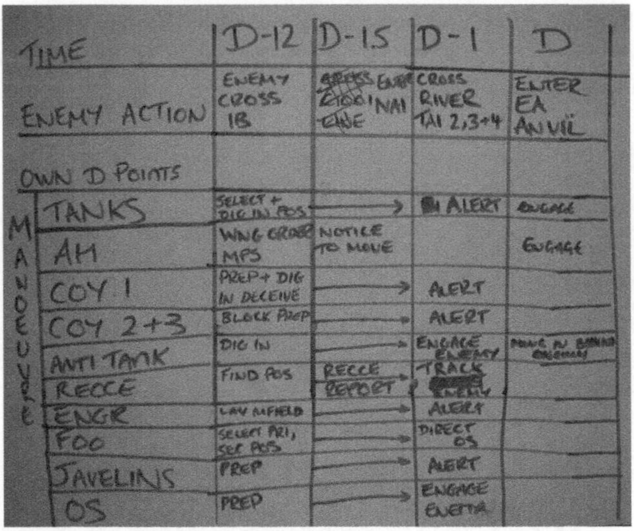

Figure 7.8   Synchronisation matrix

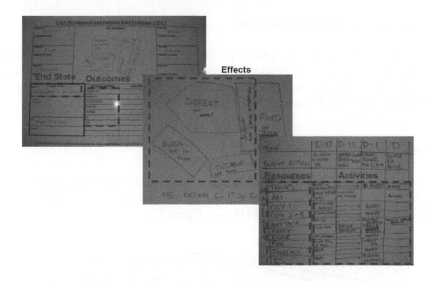

Figure 7.9   Diagram showing where content for the Abstraction Hierarchy is extracted from

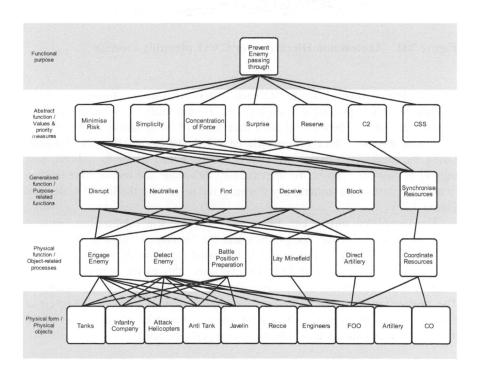

Figure 7.10   Abstraction Hierarchy for CAST planning exercise

Development of Visualisation Ideas

The previous section has introduced the concept of EBO and explored how the Abstraction Hierarchy can be used to express the relationship between physical objects and the overall end state graphically. This section uses these insights to introduce a number of concepts for visualising CoA information.

The Synthesis of EBO and CWA

Traditionally, in the combat estimate process, CoAs are evaluated based upon their individual compliance with the decision criteria listed in the Values and Priority Measures in the Abstraction Hierarchy (see Figure 7.11).

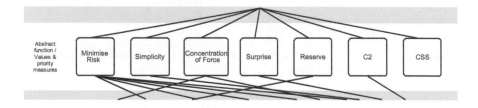

Figure 7.11   Abstraction Hierarchy for CAST planning exercise

The traditional approach is to rank each of these criteria out of three. A graphical approach to this is presented in Figure 7.12. Here a 'milk bottle' approach, that partially colours the blocks indicates their level of compliance, is proposed to represent the system's level of concordance with the Values and Priority Measures. Using the means-ends links, the effects can be traced down to indicate how individual measures can be manipulated. By 'hovering' the mouse over the box in question, links are automatically shown in red to indicate the linked nodes (see Figure 7.12).

The completed model can be used to explore the relationship between adjustments to individual components and the resultant change to the overall end state. The fidelity of this model is increased significantly by also applying a weighting to each of these links to give some idea of the magnitude of the reaction they could have; in the same way as the decision matrix in Figure 7.6. The Abstraction Hierarchy represents the same system at a number of levels of abstraction. As the previous chapter has shown, the nodes at each level of the hierarchy can be evaluated to indicate system performance.

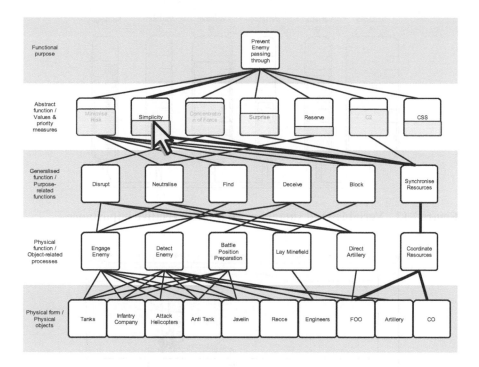

Figure 7.12 Graphical representation indicating the milk bottle representation and the highlighting of links

Display Ideas

The concept of 'milk bottles' can be extended to add in a summary 'milk bottle' (see Figure 7.13). This summary shows the overall change to the systems functional purpose. Here a link is shown to indicate that the collective values can be summarised in a bar at the end. When the variables are of differing importance, the width of the tubes can be altered to reflect this (see Figure 7.13). In this example, variable A is of very high importance, twice that of variable B. To reflect this, the width of the tubes is twice the width of variable B. this fluid based system uses a 'metaphoric reference' (see Chapter 3) based on simple fluid dynamics to allow the user to develop a mental model of how the system is functioning.

Using slider bars, values at the bottom of the hierarchy can be manipulated to indicate how well actions are performing. The resultant changes to 'actions' has a real-time impact on the 'effects' and in turn the 'outcomes'. The representation can be simplified by removing unnecessary links. By selecting a node within the diagram of interest, the unconnected nodes can be suppressed (see Figure 7.14).

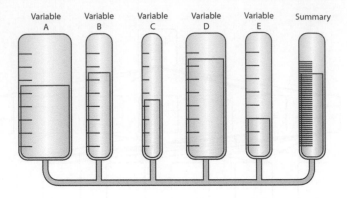

Figure 7.13   Variables and summary values with relative importance

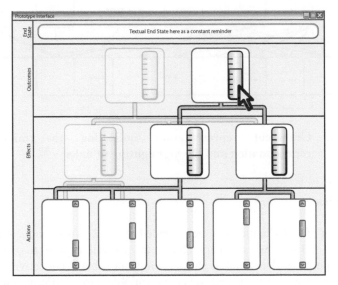

Figure 7.14   Diagram with unused nodes suppressed

In order to compare two or more CoAs, it is advantageous to view them within the same representation. This could be done by modifying the display shown in Figure 7.14 to show two CoAs in two different colours (see Figure 7.15).

The, rather abstract, choice of a fluid dynamics metaphor is not necessarily the optimum representation type for this information. By exploring alternative representation techniques and evaluating them experimentally, design recommendations can be extracted.

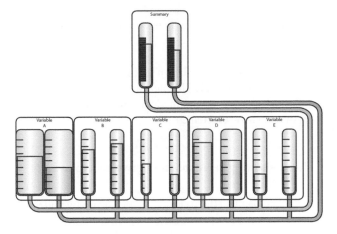

Figure 7.15 Comparison of two CoAs

Experiments

In order to establish design recommendations a series of interfaces were developed. These developed interfaces were then evaluated via a set of controlled experiments were constructed in two phases. The first of these phases required participants to make decisions based upon the information displayed in previously generated (static) interfaces. The second phase of experimentation, required participants to use a dynamic interface to make a decision based on a simple route selection paradigm.

Phase 1 – Interpretation of Existing Displays

Design

Phase 1 of the usability study was created to explore human performance issues using static CoA display stimuli. The task of the participant was to interpret the display and select an 'optimum' (i.e. best) CoA. The within subjects factor (display type) had six levels: a numeric ranking display, process comparison display, decision matrix display, tubes display and gears and locks display. The displays were developed to address a number of key questions:

- Do participants perform better with displays based upon 'bar' or 'pie' charts?
- Do participants perform better with textual or graphical displays?
- Do participants perform better with exact numerical entry or rough order of magnitude slider-bars?

- Do participants perform better when the effects are grouped by outcome or grouped by hierarchy?

Illustrations of the display types are shown in Table 7.2 with enlargements shown below in Figure 7.16 to Figure 7.21. The measured dependent variables were, time to make decision and select an optimum CoA; and a binary measure of accuracy (correct CoA/incorrect CoA selected).

Table 7.1   Table showing relationships between displays

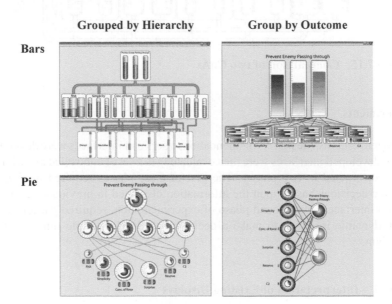

Table 7.2   Table showing relationships between displays

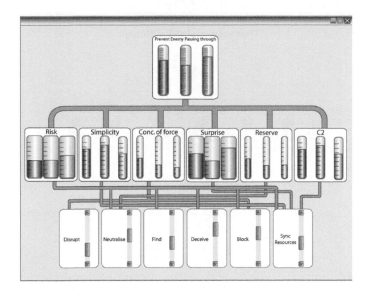

Figure 7.16 Process control (tubes) concept

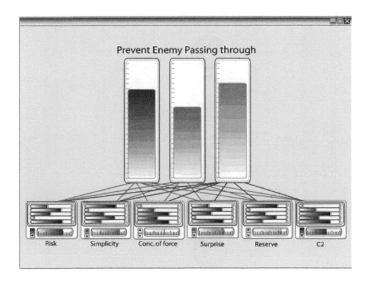

Figure 7.17 Process comparison concept

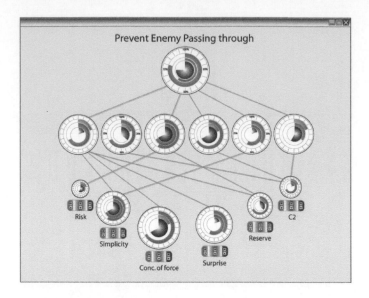

Figure 7.18   Gears concept

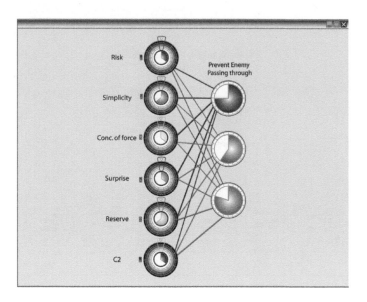

Figure 7.19   Combination lock concept

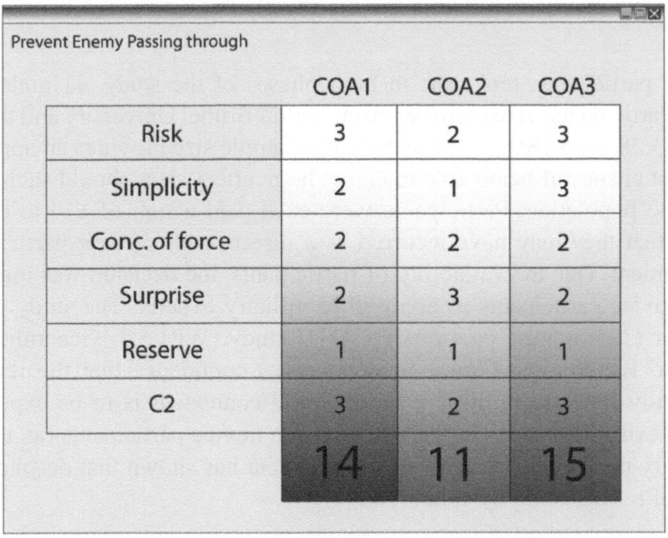

Prevent Enemy Passing through
COA1
COA2
COA3
Risk 3 2 3
Simplicity 2 1 3
Conc. of force 2 2 2
Surprise 2 3 2
Reserve 1 1 1
C2 3 2 3
14 11 15

Figure 7.20 Current (base) system

Risk 1
Simplicity 3
Conc. of Force 2
Surprise 4
Reserve 3
C2 2
COA 1
3 9 4 6 6 2
3 3 2 2 2 1
75
COA 2
1 6 0 4 6 4
1 2 0 2 2 3
60
COA 3
3 6 4 8 9 5
2 2 2 1 3 2
80

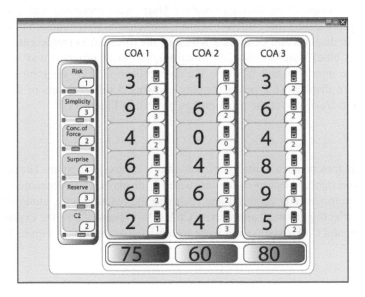

Figure 7.21 Decision matrix concept

Participants

Fifty-nine participants took part in both phases of the study, 41 males and 18 females. Participants were recruited from within Brunel University and their mean age was 23.28 years (SD = 7.29 years). The sample size provides an approximate 86 per cent chance of being able to detect large effect sizes should such an effect exist in the population. Participants were each paid a sum of £10 to cover any expenses that they may have incurred as a direct result of their participation in the experiment. Due to availability of participants, the decision was made to use civilian novice participants as apposed to military experts. The study published by Grether (1949) and a previous HFIDTC study (WP1.5.3 'Generalising from Novices to Experts in Military Studies) gives confidence that the results will meaningfully relate to military personnel and context. It is to be expected that there will exist differences in 'extent' between novice participants (as used here) and military personnel. The previous experiment has shown that despite this, the 'pattern' of results is highly concordant.

Experimental Methods

The first stage of experimentation involved the evaluation of the existing displays generated because of the literature review. Users were presented with a series of displays representing three different CoAs. Participants were required to interpret these static displays and determine which of the three CoAs (represented as red, green and blue) were optimal. A 'Macromedia Flash' program was written to display the interfaces and facilitate the data collection. This program presented 30 different displays (6 types of display showing 5 different sets of data). The onscreen instructions for the program can be seen in Figure 7.22

Equipment

This study took place in a small environmentally controllable room. The programs were run on standard laptop-computers with 15" displays. The 'Macromedia Flash' program automatically captured and recorded the reaction time, along with any errors in selecting the most appropriate CoA. Illustrations of how the experimental procedure and CoA visualisation were instantiated are shown in Figure 7.22 and Figure 7.23.

Procedure

The six interfaces were presented independently in the same way. At the start of the experiment participants were presented with a screen, containing instructions (see Figure 7.22). Participants were required to select the display which they believed to be optimal by clicking on the respective coloured button (circles at the bottom of the display; see Figure 7.23); data was recorded on the time taken to make

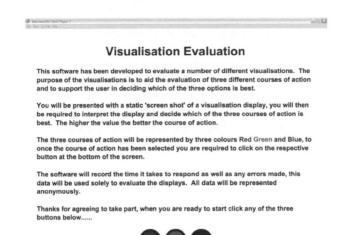

Figure 7.22   On screen instructions for software

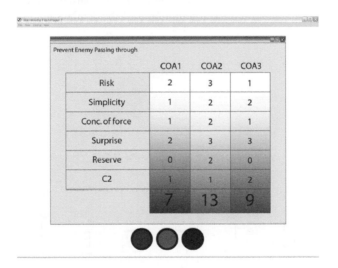

Figure 7.23   Screen shot of software

this decision along with any errors made (incorrect button pressed). Data was always presented in the same order in each of the displays (from left to right red, green, blue). The 30 displays (6 interfaces; 5 data sets) were presented randomly to remove any order effect.

1. Participants were welcomed and introduced to main aims of the trial.
2. Demographic information was collected (age, gender).
3. Participants were asked to complete a short practice program for a unique graphical display.
4. The participants were offered the opportunity to pose further questions.
5. The participants tackled each of the graphical solutions in a random order.
6. After each test was completed, the results were recorded.

Results

Participants were asked to look at a static display, interpret it and decide on a course of action. The speed of their decision and its accuracy were measured.

Time

Figure 7.24 shows each of the displays plotted along the x-axis and the mean time that it took participants to respond along the y-axis. The results for each display are shown on a box plot; here the thick black line shows the mean response rate with the white box showing the results within one standard deviation. In general, there seems to be little to separate the majority of the displays with the results for

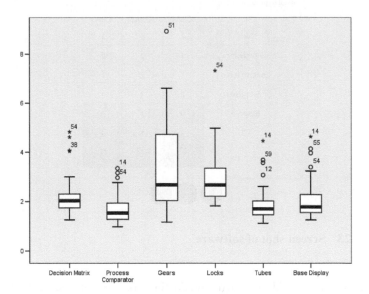

Figure 7.24   Boxplot showing the mean and associated variance in the time taken for participants to select a CoA from each one of the static display type

the decision matrix; process comparator; tubes and base display appearing very similar. The results for the lock display and the gears display indicate that, for some participants, these displays were more confusing. Within this general pattern of findings, it can be observed that there are a number of outliers and even some extreme values. The variation in response time indicates individual differences in ability between the participants (some individuals take considerably longer than others, despite the same display type being used). In terms of analysis, this suggests a degree of statistical caution/conservatism might be warranted.

The mean performance times obtained within each of the six CoA display types differs significantly at the 1 per cent level: $F(1.76, 103.69) = 57.55$; $p < 0.01$. This significant finding is accompanied by a very large effect size; partial eta squared $= 0.49$ suggesting that, not only is this a phenomenon that can be generalised from the current study to a wider population, but, that it is a large and meaningful phenomena from a design as well as human performance point of view.

Having detected the presence of a main effect extant between the six conditions, the question turns to consider precisely what display types differ significantly from each other. A somewhat conservative post-hoc technique (Bonferroni) was employed in order to undertake an exhaustive series of pairwise comparisons among the six displays to form the basis for a rank order. Every such comparison was statistically significant at the 1 per cent level, except, that is, for the gear and lock displays (which were not significantly different to each other in terms of performance time; $p = ns$). This does not connote that there is literally zero difference between the two conditions (indeed, Figure 7.24 above might suggest that the lock display is somewhat quicker). This finding can be interpreted as being of considerably less magnitude (and importance) compared to the others; in other words, from a practical point of view a design intervention based on the lock and gear displays is not likely to yield a particularly meaningful change in performance time. Figure 7.25 is a profile plot that provides a graphical summary of the post hoc analysis.

In summary, the pattern of results obtained presents a fortuitous case in which Figure 7.24 can be interpreted both literally and statistically. As such, the Process Comparator display reveals itself as being (literally and statistically) the quickest in terms of CoA selection time, this is closely followed by the Tubes and the remaining displays to follow in the following order: Base; Decision Matrix; Locks and then Gears (with the difference between the locks and gears not being statistically significant).

Accuracy

As well as performance time, performance accuracy was also measured using a dichotomous measure (1 = Correct CoA selected/0 = Incorrect CoA selected). Figure 7.26 shows the proportion of correct responses given by the participants within each display type/condition. The results appear, visually at least, to be conclusive. They show, firstly, that with each of the five separate iterations

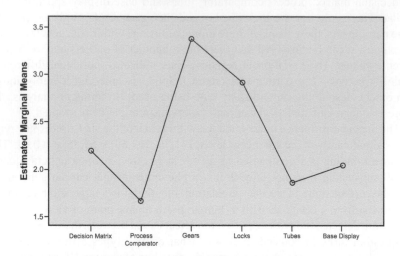

Figure 7.25   Profile plot illustrating the pattern of results derived from a post-hoc analysis of the significant main effect (p < 0.01) achieved when comparing the six CoA display types

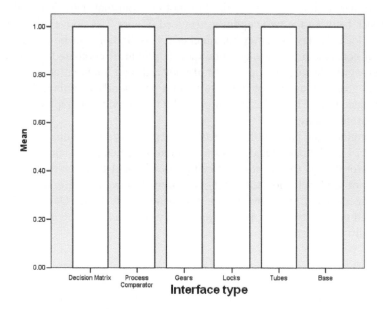

Figure 7.26   Bar chart showing that all conditions (bar the Gear display) are able to reach a ceiling level of 100 per cent accuracy (the Gear display reached a mean of 95 per cent)

within which a participant gave a response that could be assessed for accuracy, they achieved a ceiling level of 100 per cent. The key point seems to be that all display types support very high levels of accuracy. Secondly, and despite this, the Gears display dipped slightly (if markedly) from the rest (from 100 per cent to 95 per cent). Indeed, a Cochran's test shows that this finding has some statistical foundation: Cochran $Q = 15.0$; df $= 5$; $p = 0.01$. The slight dip in accuracy for the Gears display has arisen not merely from random error, and can be inferred to the wider population.

Unfortunately, this data is not particularly amenable to formal statistical power tests, which necessitates a more descriptive interpretation. It suggests the following:

- All display types seem to support very high accuracy levels (notwithstanding some aggregation of the data to present a mean for the group, accuracy reaches the ceiling level of 100 per cent).
- A probability/cost/time trade-off will determine whether the small accuracy decrement for the Gears display is meaningful from a display design perspective.

Time Accuracy Trade Off

Bearing in mind that all displays hit the ceiling level of accuracy the dilemma faced with how to trade off time and accuracy in this case seems simple. It would seem completely reasonable to base the selection based entirely upon the time taken to respond to the displays. Based upon the large number of displays to choose from it may be suitable to reject the Gears concept at this stage as a results of its slower reaction time and it statistically significant worse error rate. The rank ordering of the displays based upon the first phase of experiments (with the best display first) would be:

1. Process Comparison (see Figure 7.17).
2. Tubes (see Figure 7.16).
3. Base (see Figure 7.20).
4. Decision Matrix (see Figure 7.21).
5. Locks (see Figure 7.19).
6. Gears (see Figure 7.18).

Phase 2 – Ease of Generation and Usability of Displays

Design

Phase 2 of the usability study was created to explore the human performance issues relating to dynamic, rather than static CoA display stimuli. The task of the participant was to plan a route through the centre of London by selecting from three possible alternatives. The criteria for selection/interpretation related to; speed, distance, traffic measures and safety. The weightings the individual factors had on higher levels of abstraction were consistent and fixed. The within subjects factor (display type) was based upon the displays used in Phase 1. Because of the changes to the scenario, the displays in turn changed; due to technological constraints (within Microsoft Excel), it was not possible to model the 'Combination Lock' concept dynamically. Illustrations of the display types are shown below. There were four measured dependent variables:

1. Time to make decision and select an optimum CoA.
2. A concordance measure based on a comparison between the CoA recommended by the display and the one which the participant would likely choose in practice.
3. The participant's subjective rating of the ease of use and.
4. The participant's subjective rating of confidence with the decision made.

Participants

The same participant sample was used in Phase 2, as was used in Phase 1; Phase 2 simply followed directly on from Phase 1.

Experimental Methods

The second stage of testing aimed to investigate how intuitive to use the displays were. For this phase of the experimental process, participants were presented with dynamic interfaces generated in Microsoft Excel. These displays could be manipulated by means of textual data entry, slider bars and buttons.

In order to evaluate the displays a scenario was sought that was similar to a military context, which did not require any previous training. A simple route selection exercise was selected. This scenario involved the participants being presented with a paper map showing three similar short routes through the city of London each starting at the same point and ending at the same location (see Figure 7.27). Participants were required to compare these routes on a number of factors including:

1. Predicted average speed.
2. Estimation of the distance travelled.
3. Number of traffic measures (traffic light, roundabouts, junctions).
4. Interpretation of the risk of accident.

The participants were helped in the decision making process by a dynamic interface on a standard laptop. The interfaces were manipulated by numerical inputs, button clicks or the use of slider bars. The ratings of these options influenced a computer generated rating of the quickest and safest route; this in turn influenced the computer rating of optimal route rating. An example of one of the maps can be seen in Figure 7.27. The displays evaluated can be seen in Figure 7.28 to Figure 7.32.

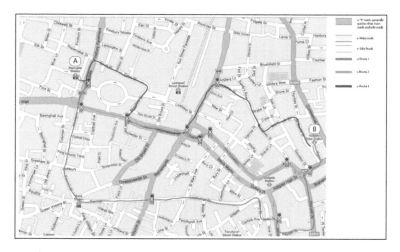

Figure 7.27   Example of routes selection card

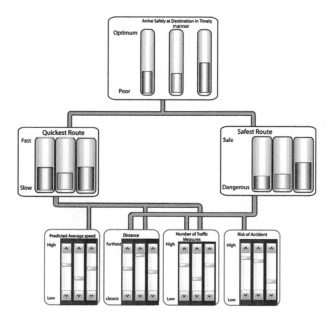

Figure 7.28   Process control (tubes) concept

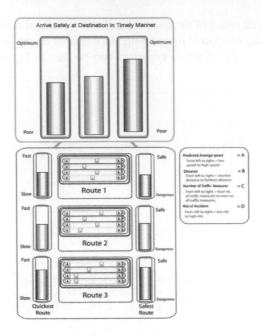

Figure 7.29   Process comparison concept

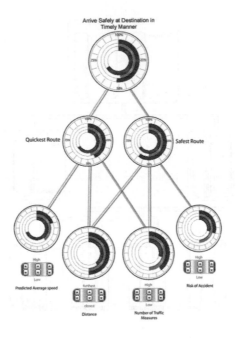

Figure 7.30   Gears concept

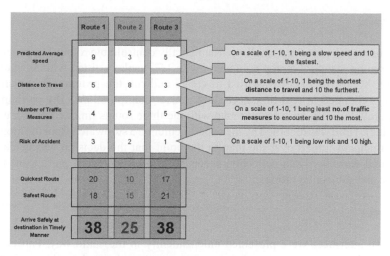

Figure 7.31 Adaptation of current (base) system

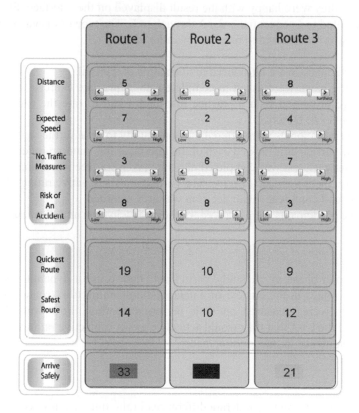

Figure 7.32 Decision matrix concept

Equipment

This study took place in a small environmentally controllable room. The programs were run on standard laptop computers with 15" displays, fitted with an external mouse. Microsoft Excel was used to create and run the program in which the displays changed in real time according to user manipulations. The participants were provided with an A4 paper map with the three routes (marked in red, green and blue). Time was recorded by the experimenter on a stopwatch from the moment the participant turned over the map to the point that the decision was made and delivered verbally. Data collection sheets were used to record participant time, choice and subjective ratings.

Procedure

The five interfaces were presented independently in the same way. Participants were required to study a map showing three routes and rate the route against the experimenter-defined criteria. Participants were allowed to re-adjust their chosen rating until they were happy with the result displayed on the interface. Five maps were presented in a sequential order A through to E. The type of display used with each map was randomised to remove any order effect.

1. Participants were introduced to main aims of the trial.
2. Participants were asked to complete a short practice exercise where they were presented with a practice map and asked to rate the route upon a number of factors.
3. The participants were offered the opportunity to pose further questions.
4. The participants tackled each of the interfaces in a random order.
5. After each test was completed, the time taken was recorded along with the participant's choice of route. The participant's subjective rating of ease of use and confidence with decision was also recorded.
6. The participants were thanked for their participation and dismissed.

Results

In this second phase of the study, participants had to use a dynamic version of the CoA displays in the context of a naturalistic planning task (route planning). Participants were measured on their performance time, how well the CoA represented the CoA that they would actually choose, subjective ratings of ease of use and confidence in the decision made.

Time

There appear to be some moderate differences in the time taken to complete the CoA task contingent upon the dynamic display type used. Figure 7.33 seems to suggest that the tubes concept is performing better than the other displays, as its mean response time (the thick black line) appears to be lower.

The mean of the times taken to complete the CoA task for each of the five display types differed significantly at the 1 per cent level, as revealed by a repeated measures ANOVA: $F(4, 224) = 10.42$; $p < 0.01$. This finding was accompanied by a large effect size: partial eta squared $= 0.16$. This suggests that such differences as do arise in performance time between the display types are meaningful from a design and human performance point of view.

Post-hoc tests using the Bonferroni procedure (to control for Type I errors) show that no significant differences exist at either the 1, 5 or 10 per cent level between the Base display and the Process Comparator display. Likewise, no significant differences exist between the Decision Matrix, the Tube display and the Lock display ($p = ns$). The profile plot shown in Figure 7.34 conveys what seems to be happening within the data.

The post-hoc tests, in terms of significant differences at the 1 and 5 per cent level, communicates that the Base display and Process Comparator displays are significantly different from the remaining displays; the former resides on a high plateau, the latter on a lower plateau. In summary, it seems that the Decision Matrix, Tube display and gear display are significantly quicker than the Base and Process Comparator displays.

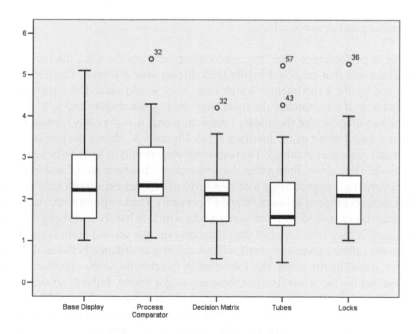

Figure 7.33   Boxplot showing the mean times taken (with associated variance) to complete the CoA task with the dynamic display types

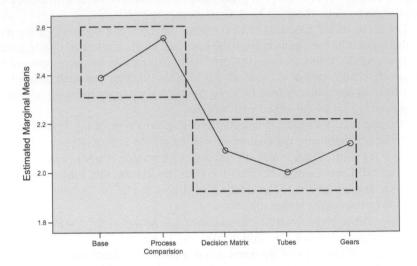

Figure 7.34 Profile plot illustrating the relationship between the different display types in terms of performance time

Suggested versus Actual CoA

As well as performance time, the concordance between the CoA the participant would take and that proposed by the CoA display was assessed. The participant was asked by the experimenter which route they would take. The experimenter recorded a '1' if this matched the route suggested by the display and a '0' if it did not (the formulas behind the displays were the same; therefore any change in this value is a direct result of the interface type). Figure 7.35 shows the proportion of concordant responses obtained. The results appear, visually at least, to be relatively persuasive. They show, firstly, that the concordance between actual and obtained CoA is very high, approaching a ceiling level of 100 per cent (with only a small within subjects range of between 90 and 97 per cent). The key point seems to be that all display types accord with the user's CoA; which is hardly surprising in some senses as it is they who supplied the data; however, the second point is that there are distinct (albeit) relatively small differences in concordance between different displays. Based on the graph, the Tube display matches the users expectancy best, the Base display the worst; but hardly to any major extent. Indeed, no significant main effect of display type on concordance was in evidence: $X2 = 3.52$; $df = 4$; $p = $ ns. It could be hypothesised that this small change could be a result of the graphical feedback supplied by the Tubes display allowing the users to more easily tweak the display to their required output.

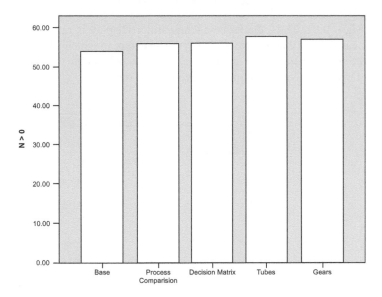

Figure 7.35 **Bar chart showing the extent of concordance between participant's own CoA and that proposed by each of the five display types**

Confidence in Decision

Participants were asked to rate their confidence in their decision (based on what the CoA display was recommending). The participant verbally rated the strength of their feeling on a five point Likert scale with the options: very unconfident; unconfident; neutral; confident and very confident. At the descriptive level there were literally no differences (hence no graph), the median ratings received by all display types being M = 4 ('confident'). Statistically no effect of meaningful size was detected at the 1, 5 or 10 per cent levels: X2 = 6.58; df = 4; p = ns.

Ease of Use

Participants provided a self-rating as to the 'ease of use' of the CoA display in relation to the route-finding task. The participant verbally rated the strength of their feeling on a five point Likert scale with the options being: very hard; hard; neutral; easy and very easy to use (see Figure 7.36). The median of those scores for each display type seems not to suggest any difference in strength of feeling; all apart from the process comparator display are rated as 'easy' (not 'very easy'). The Process Comparator Display is rated nearer to the neutral mid-point of the scale compared to the others. This result is statistically significant A Friedman test detected a main effect consistent with Figure 7.36 below: X2 = 32.83; df = 4;

Asymp. Sig < 0.01. Thus, the Process Comparator Display is rated significantly lower for ease of use than the other displays. The extent to which a 0.5 drop in the relative rating is meaningful in any practical sense is debateable (and presumably consistent upon the overall picture conveyed by the results).

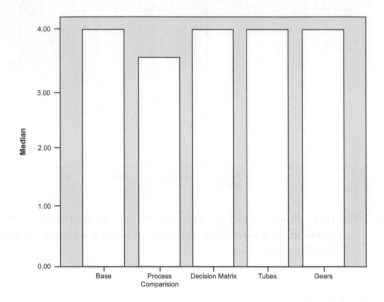

Figure 7.36   Bar graph showing the median ease of use ratings given in relation to each of the five display types

Tradeoffs Between the Dependent Variables

Each of the results captured can be summarised in Table 7.3. Here exact values have been replaced by the statistically significant differences. The time data has been broken down into two statistically significant groups (slow and fast). Concordance shows all of the displays at 'Very High', with the exception of the Base display which showed a very small, but statistically significant, reduction; marked as 'High'. Confidence has been marked as 'confident' for each of the displays as there were no statistically significant defences between the displays. For ease of use, each of the displays has been recorded as 'Easy' with the exception of the Process Comparison, which showed a statistically significant degradation on the subjective ratings. Table 7.3 would suggest that when the usability is considered there are three optimum displays, the Tubes display, the Decision Matrix and the Gears display, with all three receiving the highest ratings in all of the sections. The Process Comparison and the Base display both received statistically significant worse results in two of the sections

Table 7.3   Summary of results for phase 2 by display

	Display	Time	Concordance	Confidence	Ease of Use
Base Display		Slow	High	Confident	Easy
Process Comparator		Slow	Very High	Confident	Neutral
Decision Matrix		Quick	Very High	Confident	Easy
Tubes		Quick	Very High	Confident	Easy

Table 7.3   *Concluded*

Display		Time	Concordance	Confidence	Ease of use
Gears		Quick	Very High	Confident	Easy

Experimental Conclusions

When the phase 2 results are compared with the rank ordering based on phase 1, further conclusions can be drawn. By synthesising the results of both experiments, a rough order can be established (as shown in Table 7.4). Here the optimum solution can be seen to be the 'Tubes' display closely followed by the 'Decision Matrix' display. Both of these displays scored maximum ratings in phase 2 and scored highly in phase 1. The Process Comparator ranked third, as this scored highest in the phase 1 experiment, but it was found to be more difficult to use dynamically in the second phase of experiment showing a significantly slower time for use and also receiving a statistically significant worse subjective rating of ease of use. The Base display derived from the paper process used at CAST ranked in fourth. This display came in fourth position in the first phase and received a statistically slower time in the second phase of experimentation. It is hypothesised that these slower results are a product of the textual nature of the display. The Gears and the Lock displays are ranked fifth and sixth due to their poor performance in the first phase of experiments. It was found that these attempts to compact the data into a small area were perceived as convoluted and difficult to interpret.

Implications for Design

Some broad conclusions can be drawn from the way that the different displays performed under participant evaluation. Table 7.4 shows the displays placed in order based upon their compatibility with the experimental variables.

Table 7.4   Summary of results ordered

	Display	Order (Ph 1)	Time (Ph 2)	Concordance	Confidence	Ease of Use
Tubes		2	Quick	Very High	Confident	Easy
Decision Matrix		4	Quick	Very High	Confident	Easy
Process Comparator		1	Slow	Very High	Confident	Neutral
Base Display		3	Slow	High	Confident	Easy
Gears		6	Quick	Very High	Confident	Easy
Locks		5	N/A	N/A	N/A	N/A

The Gears and Locks displays were both received significantly slower reaction times in the first phase of the experiment. It is postulated that any deviation from widely accepted conventions (bars, graphs and numbers) need to be considered carefully. Participants reacted significantly slower to displays with pie chart based interfaces. This view is supported by Tufte (1983), who points out that:

> Tables are preferable to graphics for many small data sets. A table is nearly always better than a dumb pie chart; the only thing worse than a pie chart is several of them, for then the viewer is asked to compare quantities located in spatial disarray both within and between pies... Given their low data-density and failure to order numbers along a visual dimension, pie charts should never be used. (Tufte 1983)

Extreme care should therefore be taken when aggregating large amounts of information into bespoke displays.

For the task of rough order of magnitude planning (using approximate values and outputs to understand qualitative differences between the systems), the results would indicate that participants favoured slider bars over actual textually entered numerical results. It is conceived that the spatial difference between the two points is more rapidly interpreted than abstract numerical values. This judgement can be inferred from the similarities between the Base display and the Decision Matrix concept; with their main difference being the data entry method. Care is required in extending this generalisation beyond this scenario; in situations where exact numerical results are common a textual input is likely to be favoured.

An interesting observation can be made from the difference between the Process Comparator display and the Tubes display; these two displays are very similar in their concepts both depending heavily on bar charts and slider bars. Their main difference is in the way that the data is organised. The Process Comparator display grouped the data tightly so that each CoA was compared in turn rather than grouping by the variable as with the Tubes display. It is conceived that by grouping by variable it was easier for the participants to compare each CoA on individual factors directly. Another key factor is that due to the way the data was structured in the Process Comparison display it was less clear what the limits represented on the slider bar.

Chapter Summary

This chapter has explored how the CWA framework can inform the design of a military decision support tool. The EBO terminology was mapped on the Abstraction Hierarchy and the benefits of structural means-ends links explored. Building on the approach introduced in Chapter 5 a rating for each node in the hierarchy was applied; the ratings for higher levels of abstraction, through tracing the means-ends links, were approximated automatically by manipulations to the lowest level of the hierarchy. This, theoretically grounded, approach to the decision support tool formed the basis for a number of interfaces. A number of key questions were addressed on the optimum interface type. These displays were evaluated in a controlled experimental environment to draw broad design recommendations.

The statistical analysis performed enables the results to be inferred to a wider population. It is important to reiterate that the participants for this experiment were drawn from a civilian population. However, confidence is gained from previous, similar, experiments (Walker et al, 2006b; Grether, 1949) which found that, although absolute values may differ between military and civilian groups, overall trends remain the same. A previous HFIDTC study (Walker et al, 2006b) investigated participant's ability to conduct Battlespace Area Evaluation. This study also investigated the performance differences between civilians and military personnel. The results found that, although differences in 'extent' (or magnitude) between novice participants and military personnel existed, the overall trends remained; thus, concluding that it was applicable to use civilians to identify performance differences between support tools. Grether (1949) conducted a similar experiment evaluating a number of different aircraft altimeter displays. He conducted the same experiment with both USAF pilots and collegemen. The results of this experiment identified that, whilst the collegemen took longer to read the displays, the rank ordering of the displays, based upon performance, remained the same. Based upon these results, Grether (1949) suggests that previous experience of the subjects is of relatively minor importance in an experiment of this type.

The experimental results from this study have identified the performance benefits of digital displays that support Rough Order of Magnitude (ROM) planning. These displays, developed in standard applications such as Microsoft Excel, can be generated quickly and allow real time feedback to the user on the implications of changing activities or effects further down the hierarchy. It is conceived that these tools may be of some benefit in evaluating CoA situations where the relationship between actions, effects, decisive conditions and end-states is understood and relatively static. It is also conceived that it may also be possible to automate some of the processes currently carried out manually; feeds based upon from other data collection systems could potentially be used to populate the display.

It should be noted that this study was conducted as a small scoping study; the remit was not to produce a finished product; rather, the aim was to investigate, in detail, the factors that may influence the success of a chosen CoA decision support tool. Care needs to be taken in the interpretation of these results. Decision-making is, inherently, a cognitive process, informed by complex understanding of the environment and based heavily on experience. In short, Commanders make decisions, not computers. The role of a 'decision support tool' should not be to take on the role of decision making, instead the role should be to provide timely information in a format that enhances the Commander's decision making performance. Any decision support tool needs to be carefully considered, it should be extensively compared to a paper-based benchmark examining any changes in macrocognitive performance, such as; decision making, sense-making, planning, adaptation, problem detection and coordination

Methodologically, this chapter has shown that CWA provides a clear structure for the design process. The approach, prescribed in the Abstraction Hierarchy, of viewing the same system at different levels of abstraction has been exploited. Further, the approach of using each level to evaluate the system has been further exploited, along with the information contained within the structural means-ends relationships.

Chapter 8
Development of a CWA Software Tool

Introduction

CWA is currently a paper-based process; analysts frequently using the framework often use graphical software tools such as 'Microsoft Visio' to support the creation of diagrams and representations for each of the phases. The interdependencies and linkages between CWA's component phases cannot be modelled or exploited by this current setup. Given the flexibility of the framework's application, it is very difficult to supply novices with prescriptive instruction, it would be beneficial if a system could be developed that could provide the user with appropriate context dependant guidance. The software tool developed by the HFI DTC (Jenkins et al, 2007a) aims to address these two short falls of the current CWA system for the tasks of application and tuition. This chapter introduces the CWA software tool, explains the reason for its development and discusses some of the key design decisions made. The tool is then considered against the per-based and MS Visio alternatives.

Why Do We Need a Software Tool?

The prime purpose of the CWA software tool is to assist the user in developing the large number of graphical representations that support the iterative design process. The second purpose is to explain CWA to novices, systematically guiding them through the analysis by providing them with context relevant guidance. The CWA process is often criticised for being complex and time consuming, this tool attempts to provide some level of guidance as well as to expedite the documentation process to address these concerns. The tool uses the 'sensor to effecter' paradigm (Jenkins et al, 2007c; presented in Chapter 6) as a common example to guide novices through the CWA domain and the tool's application.

What Should the Tool Do, and How Should It Do It?

By creating a series of templates, the task of generating electronic analysis products can be significantly expedited. Supporting the iterative design process is a key requirement; the terminology used in the analysis products is extremely important and frequently changes throughout an analysis. Small changes to the semantics of a node description are common. When these descriptions are reused in multiple diagrams the task of updating them is not only time consuming but also prone to errors of omission. The phases of CWA each model the same domain using different constraint types. It is often possible to part complete some of the representations with information from previous phases. Via a series of dynamic links the software tool has the potential to allow forward-feed of data allowing diagrams to be partially completed, updates to nodes can also be propagated throughout the analysis.

The introduction to this book highlighted that the approach taken by an analyst conducting a CWA is highly dependent on the research aims and the domain under analysis. Many analyses will not focus on all of the five phases; the majority of analyses place a much greater focus on the first phase, Work Domain Analysis. The type of analysis conducted is likely to depict which of the phases are used and in what order and ratio. For example, interface design such as 'Ecological Interface Design' (EID) (Burns and Hajdukiewicz, 2004) tends to use only the first phase (Work Domain Analysis). An analysis of team design or training needs is more likely to focus more heavily on Control Task Analysis and Social Organisational and Cooperation Analysis as these phases capture activity, considering its distribution amongst the system's assets. The benefit of the framework is that the CWA approach can be applied throughout the system life cycle. As Chapter 5 has shown, the approach can be used for the design, development, representation and evaluation of both conceptual and existing operational systems. It should be possible for analysts to use the framework in a non-linear process using the phases applicable in an order suited to the analysis. Each of the phases can be decomposed into a number of sections. The first section of each of the phases introduces the phase in generic terms briefly explaining the principles behind the particular level of the framework. The next section provides a worked example of the analysis at the current phase consistently using the sensor to effecter case study (Jenkins et al, in press b). Full annotated provides guidance to novices, and to an idea of the finished product. The subsequent sections allow the user to construct and develop the models and documentation required; through the use of 'mouse-overs' context specific prompts can be provided to the user.

An Abstraction Hierarchy was constructed to guide the development of the software tool; this can be seen in Figure 8.1. At the highest level the functional purposes, these functional purposes form the boundary for the analysis. These are listed as Train novices; Expedite documentation of process; and to Allow real-time generation and validation of products. The requirement for real time performance is important as it allows analysts and SMEs to work together in the development of products during a discussion.

The next stage in the process of generating the Abstraction Hierarchy was to consider the values and priority measures; these measures are used to evaluate the success of the tool in meeting its functional purposes. These include Ease of assimilation (how easy it is to learn how to use the tool); Level of guidance (how much, and how appropriate the help offered to the user is); Prevent fundamental errors (the ability of the tool to stop novices from making errors that go against the basic theory of CWA); Intuitiveness; Time taken; Ease of use; and Allow collaborative working

Moving to the bottom of the hierarchy the available physical objects listed as: Documentation and understanding of the CWA theory; Mouse over activated text box; Buttons; Tabbed bar; Mouse; Menu; Database; and Internet / LAN. The affordances of these objects were then captured in the object related processes: Text based literature; Dynamic information (flash based); Understanding of paper process; Example scenarios; Display context based text; Perform functions; Export to 'MS Word' function; Allow process to be completed non-linearly; Highlight text; Highlight links; Grab and drag objects; Settings manipulation; Store data locally; Allow distributed data transfer

The final stage of the process was to link the object related processes to the values and priority measures, via a number of purpose related functions Provide background on CWA; Explain current stage of analysis; Constrain incorrect inputs; Metaphoric layout; Provide partially completed templates; Follow Windows style guides; Pass relevant data between phases; Assist in report production; Allow flexibility to tailor outputs to analysis requirements; Allow edits to automatically propagate through the system; Indicate to the user related data; Reorder products; Allow experts to skip guidance; and to Share documents (cross location). The means-ends links within the document are then validated using the why-what-how triad. The model serves as a description of the tool's domain at a number of levels of abstraction. These nodes serve as the starting point for developing the software specification.

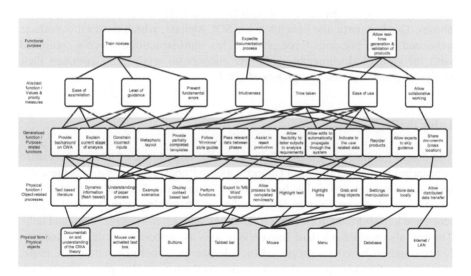

Figure 8.1   Abstraction Hierarchy for the CWA tool

The Benefits of the Tool

The dynamic nature of the software tool has many benefits over its paper-based counterpart, the tool allows files to be saved, copied, edited and rapidly transmitted. Features such as an ability to export all of the products to a 'Word document' with a single button click also make the report compiling process much simpler. The creation and revision of diagrams is simplified by the tool facilitating the ability to reorder nodes and links rapidly. The tool has been intentionally designed to be unconstrained allowing researchers to apply their own interpretations of the framework. Some functions are deliberately not permitted in order to constrain novices from making fundamental mistakes such as connecting two data processing activities in the decision-ladder, or connecting a node in the Abstraction Hierarchy to a node two levels above.

Additional benefits of the dynamic nature of the tool have also been exploited: When the user places their mouse over a node on the Abstraction Hierarchy the linked nodes both up and down the hierarchy are highlighted in red; this allows a cause and effect relationship to be examined within the different levels. The speed in which the Abstraction Hierarchy can be developed, edited and reviewed (helped by the addition of the red links) makes it feasible to carry out the generation and validation of representation from scratch with Subject Matter Experts (SMEs) in real time.

The software tool passes important data forward aiding the completion of subsequent representations. This significantly reduces the tedium of the documentation process. This also means that minor, semantic changes to text and diagram layout can be fed through from the initial stages to the subsequent phases. This has particular benefits in the SOCA phase, which reuses the products generated in the previous three phases. This semi-structured process combined with the ease of documentation further encourages analysts to continue their analysis beyond the first two phases of CWA.

Using the Tool

The design guidance in Chapter 3 discussed the importance of adopting standard approaches to interaction; Alan Cooper's axiom, 'obey standards unless there is a truly superior alternative', has been adopted in this case. The tool supports all the standard windows application functionality and flexibility with standard menus, icons and key shortcuts for common functions such as; print, save, open and help. The tool also adopts standard conventions such as supporting multiple windows. Other standard functions supported include panning, zooming, undo, cut, paste, copy, delete, font changes and spell checks. At the top are the 'phase-tabs' which can be used to quickly switch between the various phases of the framework. At the bottom right of the screen are the 'Previous' and 'Next' buttons for moving through the phases in the default order.

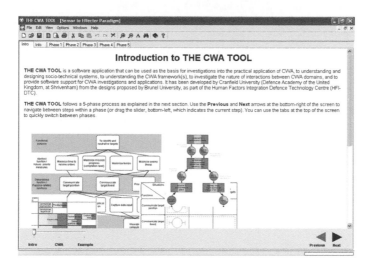

Figure 8.2   Tool introduction

In order to support collaborative working and work documentation, the tool supports a note making system. Notes can be added to all diagrams and formatted to any preference. Notes are displayed on diagrams when they are printed or exported as an image. Comments can also be added; these are only shown in the tool and when a user hovers over the placeholder.

Analysis Information

Specific details about the analysis can be added on the analysis information screen, the section is used to capture information that should be archived for other future document users. Fields exist for the following information:

- Analysis Name.
- Purpose of Analysis – the purpose or reason/aim for carrying out the analysis.
- Analysis Boundaries – boundaries for the analysis, what is (or isn't) being investigated.
- Save File Location.
- Analysts – records are made of those involved on the project; information can be stored on name, contact details and other info.
- Sources – Any sources used in the analysis can also be recorded, there is a list of the six most common source types to select from. The author, date, link and other information fields also exist.
- Glossary – A list of specialised words or terms and their corresponding descriptions can be recorded here.

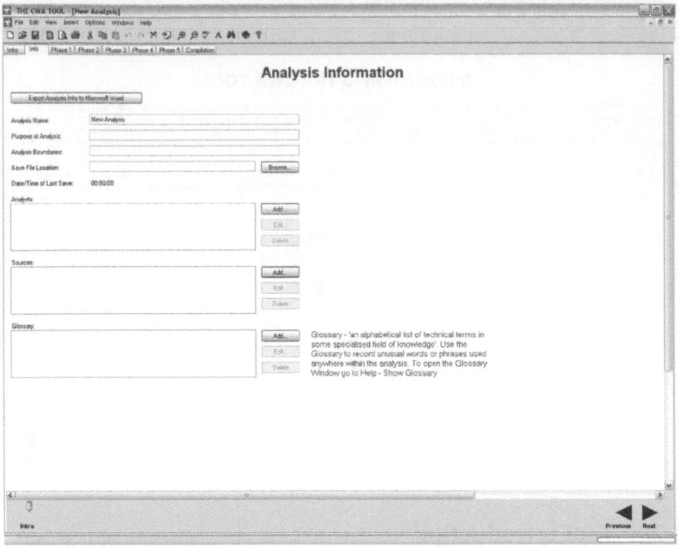

Figure 8.3   Analysis information screen

The 'Export Analysis Info to Microsoft Word' button can be used to send all of the above information to Microsoft Word in an unformatted textual list, which could form the first part of a report, for example.

Phase 1 – Work Domain Analysis (WDA)

Abstraction Hierarchy (AH)

An Abstraction Hierarchy is started by adding in nodes using the 'Add' buttons on the right hand side; the tool prompts the user with the order for adding nodes by highlighting the next row to be populated in yellow; this is advice only and the user can choose to ignore it. To edit a Node's text, the user clicks on it, or uses its right-click menu and selects 'Edit Text' (standard windows approach). This menu can also be used to access other edit options for the node; to 'clear its Means-End Links', or to 'set its Category' for the Abstraction-Decomposition Space (described below). To add a Means-End link between two Nodes in adjacent rows, the user clicks on one of the nodes and holds down the left-mouse button. A 'ghost' node appears, the user then drags this over to another node and a caption will be displayed (either Create Link or Remove Link, see Figure 8.5) if a Means-End relationship is possible, the mouse button is released to complete the action.

This drag-and-drop mechanism (a form of direct manipulation; see Chapter 3) can also be used to re-order Nodes within the same row. The links connected up and

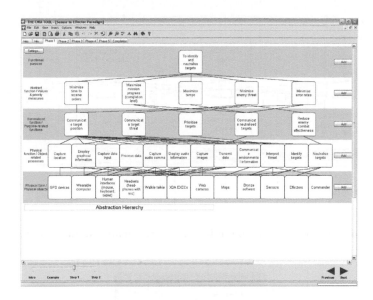

Figure 8.4   The Abstraction Hierarchy screen

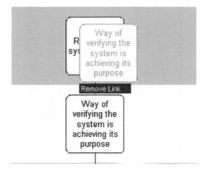

Figure 8.5   Diagram showing removing a link

down the hierarchy from a specific node can be highlighted in red by 'hovering' the mouse over the Node for a few seconds (see Figure 8.6).

The diagram can be formatted by changing the distance between the rows (see Figure 8.7).

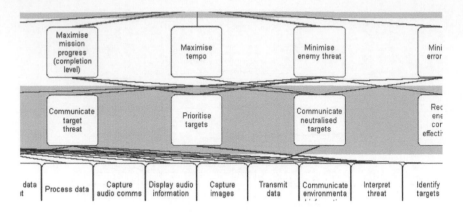

Figure 8.6   Diagram showing linked nodes to 'prioritise targets' node

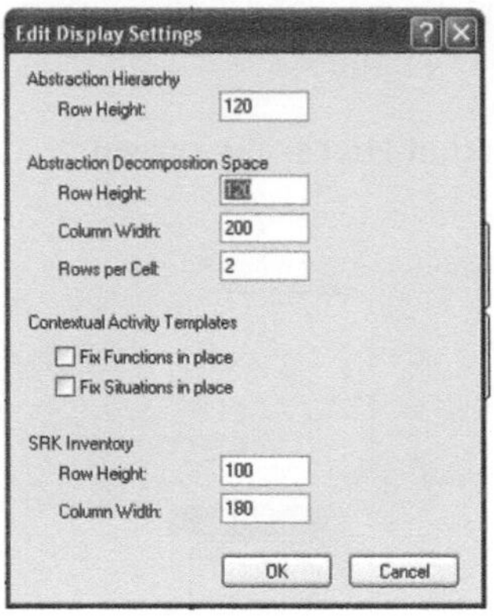

Figure 8.7   Settings display

Abstraction Decomposition Space (ADS)

Here, the Abstraction Hierarchy Nodes are allocated to levels of Decomposition, namely Total System, Subsystem and Component (these can be edited and new columns added or deleted). Nodes are simply dragged and dropped to move them into the diagram and between columns. The nodes cannot be edited within the ADS, all editing has to take place within the AH, this feature has been added to ensure that the two documents are consistent; all changes to the AH propagate through to the ADS. The diagram can be formatted to change row height and width. In addition, the rows per cell can be edited (see Figure 8.7)

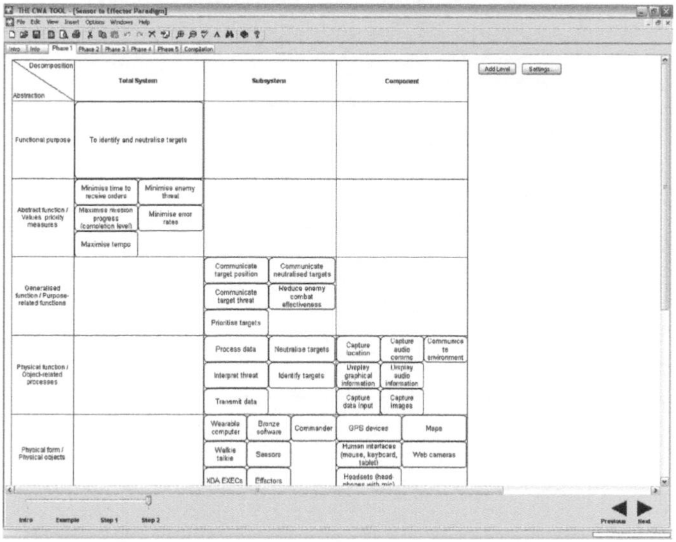

Figure 8.8   ADS screen

Phase 2 – Control Task Analysis (ConTA)

Contextual Activity Templates (CAT)

Any number of Contextual Activity Templates can be created using the 'New' button; existing templates can be accessed through the drop-down menu in the top left-hand corner.

Each template starts with only the top-left corner of a table, situations are added using the 'Add Situation' button for each Situation a new column is created in the table. Functions can be added in the same way, alternatively, functions can be imported from the AH; by clicking on the 'Import' button, a dialog box is then displayed (see Figure 8.10). This allows the user to check the boxes next to nodes they wish to import. The text in these functions can be edited, there is no dynamic link with the AH as it is expected that these functions will be edited independently.

Situations and Functions can be re-ordered using the drag-and-drop technique (see Figure 8.11).

The main facility of this version of the Contextual Activity Template is to highlight the link between Functions and Situations, namely in what Situations a Function can occur and those where it typically occurs. This is indicated by box-and-whisker diagrams, where the dotted region represents 'can' and the circle and whiskers indicate 'typical'. These notations are added using mouse clicks, by clicking on the appropriate cell the three states are cycled through (can, typically, none), as shown in Figure 8.12.

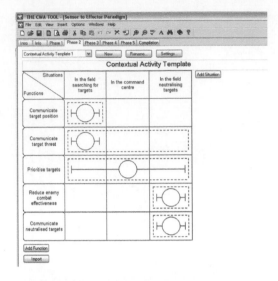

Figure 8.9   Contextual activity template screen

Figure 8.10   Import function window

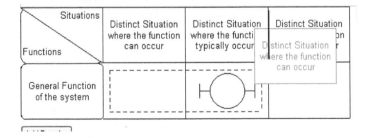

Figure 8.11   Reordering of the situations

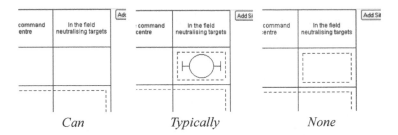

Can   Typically   None

Figure 8.12 Diagram showing the Contextual Activity Template being populated

Contextual Activity Templates (Mini Decision-ladders)

This second stage of the ConTA, supported by the tool, is an extension to the previous CAT diagrams. Where any cell whose status is either 'can' or 'typically', the option exists to complete a 'mini Decision-ladder' associated with it. The Situations and Functions cannot be edited or moved on this diagram. Any later changes in the previous step (such as to Function names) are reflected.

To highlight a component of a mini Decision-ladder, the user simply clicks on it and it turns it black; if clicked again it is un-highlighted. Ladders can be copied between cells using the drag-and-drop technique (see Figure 8.13).

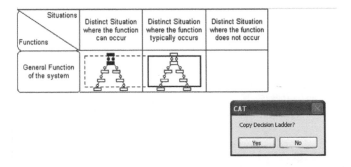

Figure 8.13 Diagram showing CAT decision-ladders being completed

Decision-ladder Diagrams

Any number of decision-ladder diagrams can be created much in the same way as the CATs. Each Ladder diagram starts with a basic template, with all the components greyed out (inactive). The components can be activated in the diagram by a mouse click, or by using the right-click menu, then Select/Deselect. Also available in this menu is an option to Show/Hide an Entry Arrow that points an arrow into the diagram at that component (the same effect can be achieved by dragging and dropping from the background to a component).

The data-processing activities (boxes) and states of knowledge (circles) within the decision-ladder can be linked (see Figure 8.15). 'Shunts' can also be made between a data-processing activity and a state of knowledge, and 'leaps' between two states of knowledge by clicking and dragging the mouse between them. If this is done on an existing link, the link will be removed. The links are constrained to stop users from adding a link between two data processing activities, as this is theoretically not possible.

It is recommended that notes are used to annotate the arrows (see Figure 8.16). Intelligent links can be made between notes and a component (one per component) by dragging-and-dropping from a component to the note. A link is created (shown in red), this is not printed or exported to an image. However, it can be used to populate cells in an SRK Inventory in the final phase automatically.

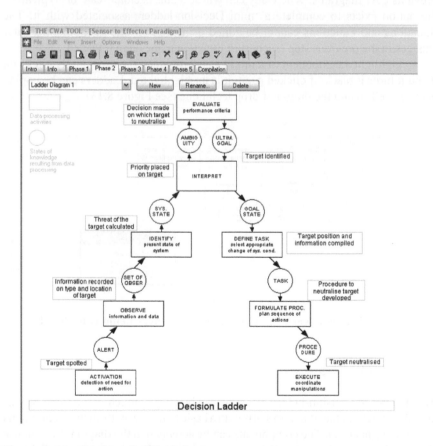

Figure 8.14   Decision-ladder screen

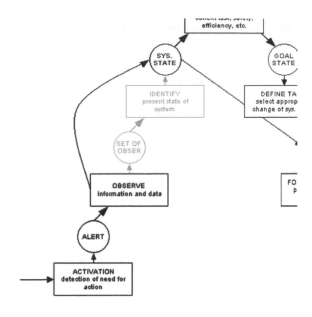

Figure 8.15   Diagram showing a linked decision-ladder

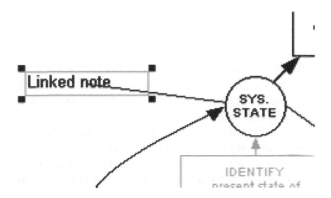

Figure 8.16   Diagram showing notes being added to part of the decision-ladder

Phase 3 – Course of Action / Strategies Analysis (StrA)

Course of Action Analysis Diagram

Any number of Course of Action analysis diagrams can be created (the term Course of Action has been adopted as this tool has been developed primarily for military application; the term strategy has a specific meaning within this domain and it was thought that this may cause confusion). The diagrams can be created and edited in the same way as the CATs.

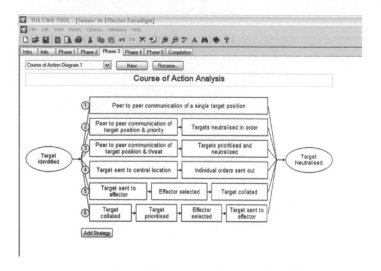

Figure 8.17   Course of action/Strategies Analysis screen

Each analysis starts with the basic start and end states, the cells between these can be edited. Additional strategies can be added between the two states to indicate another way of achieving the same end state. Additional steps can be added within each of the strategies by right clicking on an activity, the option pops up allowing the user to 'add before' or 'add after'.

Each strategy can be edited by clicking on the yellow numbered circle; these can also be re-ordered using the drag-and-drop technique (see Figure 8.18).

As new strategies are added they are numbered (shown in the shaded circles), these numbers correspond to the numbers of Decision-ladders for Strategies.

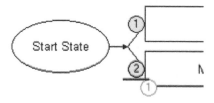

Figure 8.18   Diagram showing the reordering of strategies

Strategy Decision-ladders

A Decision-ladder can be created for every defined Strategy, the strategy diagram of choice is selected from a drop-down menu box on the left, the box on the right selects the particular strategy; the corresponding Strategy is displayed above the Ladder Diagram. The diagrams are modified in the same way as the decision-ladders in the ConTA.

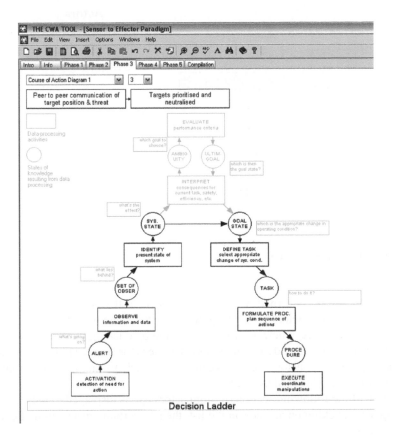

Figure 8.19   Strategy decision-ladder window

Phase 4 – Social Organisation and Cooperation Analysis (SOCA)

Actors

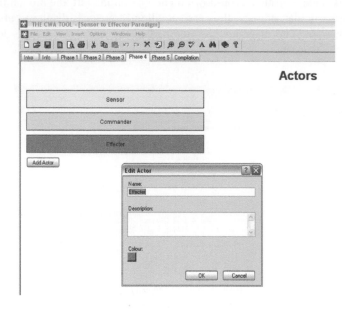

Figure 8.20   Actors screen

The SOCA phase is concerned with attributing parts of the previously generated products to particular actors or actor groups. The first stage in this process is the definition of the actors, a new actor can be created by clicking the add actor box, the name is then entered and a colour assigned to the actor, a space is also provided for a description of the actor. The actors can be reordered by dragging and dropping them, they can be edited using a right click.

In each of the following sections a 'mini actors list' is displayed to the right of the diagrams; clicking on an actor in the list, then on certain objects, will allocate the actor to them.

Abstraction Decomposition Space

Here, actors can be associated with each of the Nodes from the ADS in Phase 1. The actor is first selected from the list on the right-hand side; clicked nodes are then coded to indicate an association. Nodes can be disassociated by clicking them a second time. To change actor, the new actor type is selected from the 'mini actor list'.

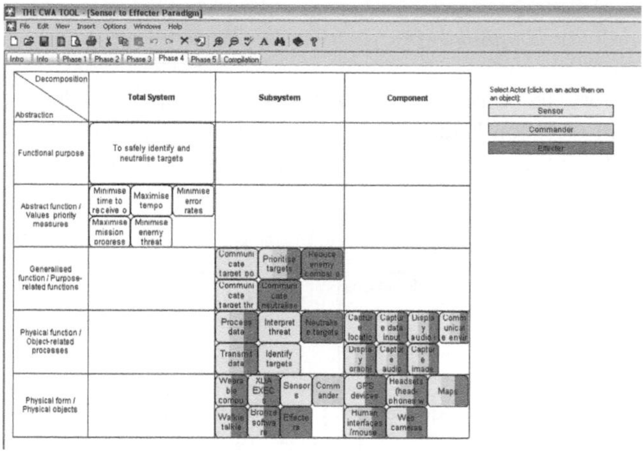

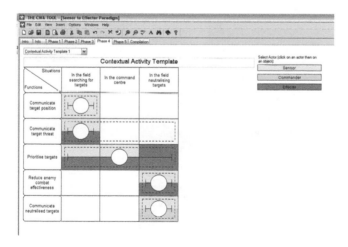

Figure 8.21   SOCA ADS screen

Contextual Activity Template

In this step, actors can be associated with any of the cells that were set as a situation where a Function 'can' occur or 'typically' occurs in Phase 2. The drop-down box is used to select from the template. The actors are attributed in the same way as for the ADS.

Figure 8.22   SOCA CAT screen

Contextual Activity Template (Mini Decision-ladders)

In this step, actors can be associated with any of the components that were selected within the mini Decision-ladders, in the CAT, of Phase 2. The drop-down box is used to select from the templates. Actors are then assigned to the individual nodes within the decision-ladder.

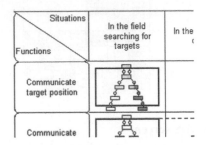

Figure 8.23   SOCA CAT Decision-ladder

Decision-ladders

Different ladders can be selected from the drop-down box, the ladders are coloured much in the same way as the other sections of the SOCA phase

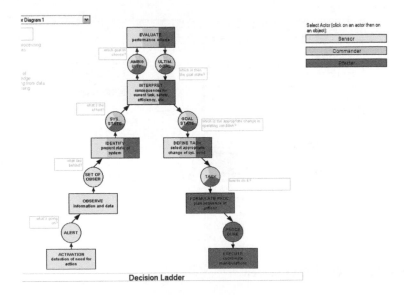

Figure 8.24   SOCA Decision-ladders

Course of Action Analysis

Different strategies can be selected from the drop-down box, the ladders are coloured much in the same way as the other sections of the SOCA phase

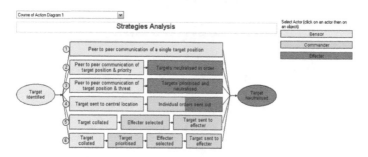

Figure 8.25   SOCA StrA

Phase 5 – Worker Competencies Analysis

Skills, Rules and Knowledge Inventory

The SRK Inventory has six columns. From left to right they are:

- A display of a mini Decision-ladder, with the respective state and activity highlighted (in black).
- Description of the 'Information Processing Step'
- Description of the 'Resultant State of Knowledge'
- A cell for entering the 'Skill Based Behaviours' which apply.
- A cell for entering the 'Rule Based Behaviours' which apply.
- A cell for entering the 'Knowledge Based Behaviours' which apply.

Any number of SRK Inventory diagrams can be created. New documents can be created using the 'New' button, and existing tables can be accessed using the drop-down menu. When a new table is added, the user is asked if they would like to associate it with an existing decision-ladder from phase 2. Associating the ladder causes the table to be partially completed with the active states along with any data from the notes section.

The text within each cell can be edited. New rows can be added using the 'add row' button, individual activity/states can be selected by colour in the decision-ladder on the left hand side of the table. The rows can be reordered by dragging and dropping the yellow numbered circles to the desired location.

It is possible to resize the table using the settings button (minimal Row Height is 80 and Column Width is 100).

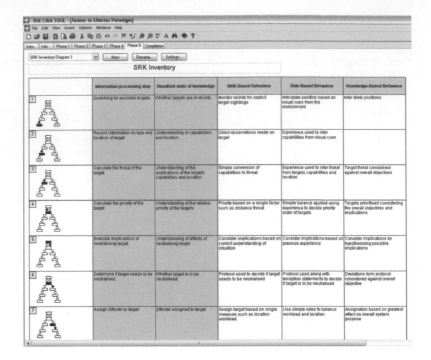

Figure 8.26   SRK table

Compilation

This final phase combines all of the diagrams from the CWA into one long list. This can then be exported to 'Microsoft Word' at the click of a single button (the 'Export All Diagrams to Microsoft Word' button at the top of the page). This functionality could be used to form the basis for a report.

Summary

One of the major benefits of the tool lies in its ability to transfer data between the phases reducing the documentation process. A summary of all of the links is shown graphically in Figure 8.27. These links are described below, the numbers relate to the numbers in the arrowheads in Figure 8.27:

1. The nodes created in the AH are copied through to the ADS where they are allocated to a level of decomposition, any subsequent changes to the AH carry forward.

2. The option exists to import the text from the nodes in the AH to the CAT. There is deliberately no dynamic link for this so subsequent changes to the AH are not reflected in the CAT.
3. The tool creates a mini decision-ladder for each cell in the matrix in which activity can take place.
4. Information for each activity can then be described in detail in a larger decision-ladder.
5. For each stage in the Strategies Analysis the option exists to create a decision-ladder, the software automatically labels the decision-ladder and places a diagram at the top showing the relevant strategy.
6. The ADS is used in the SOCA phase for coding, the link between the figures is constant so subsequent changes are reflected in the SOCA phase.
7. The CATs and decision-ladders are used in the SOCA phase for coding, the link between the figures is constant so subsequent changes are reflected in the SOCA phase.
8. The Strategies Analysis flow diagram is used in the SOCA phase for coding, the link between the figures is constant so subsequent changes are reflected in the SOCA phase.
9. Information from the notes made on the decision-ladders can be automatically imported to the WCA matrix.

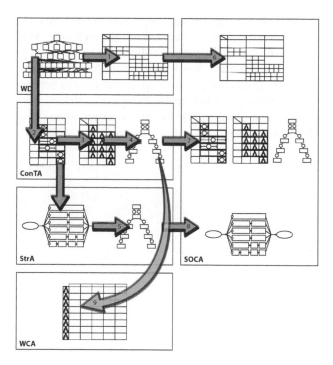

Figure 8.27   Links between phases supported by the CWA tool

Chapter Summary

This chapter has introduced the CWA software tool; the design choices made are captured using an Abstraction Hierarchy. The need for the software tool is explained along with its benefits over the paper-based approach. These are listed as the function purposes in an Abstraction Hierarchy describing the tool; Train novices; Expedite documentation of process; and to Allow real-time generation and validation of products. The training of the CWA approach to novices is supported by the inclusion of context specific guidance. The tool steps novices through the process offering hints on what to do next and the most appropriate way of approaching the analysis. The addition of the 'notes' and 'glossary' functionality allows users to add additional information for others using the same document. It is contended that the process of completing the analysis has been significantly expedited using digital templates and dynamic links. These dynamic links allow data to flow-forward partially completing other representations and making editing multiple diagrams significantly easier. The speed of the software and the ability to reorder diagrams rapidly, whilst maintaining their relationships (such as moving nodes in the Abstraction Hierarchy and keeping their means-ends links) allows the tool to be used in real-time during discussions with SMEs. Products can be generated and validated quickly and easily, further expediting the analysis process.

When compared to the paper-based approach the benefits are clear, new possibilities exist in data duplication, transmission and archiving. Arguably, a well-created template in an application such as 'Microsoft Visio' could provide many of these benefits. However, the CWA software tool has a number of additional features such as dynamic context specific help, dynamic links between phases and limiting users from making fundamentally incorrect mistakes. In the following chapter, the CWA software tool will be further evaluated to test its suitability for modelling real world domains.

Chapter 9

Does the Tool Make the CWA Process any Quicker or Easier?

Chapter Introduction

In order to validate the assumption that the CWA Tool, discussed in the previous chapter, would support the analyst; a complex military scenario was sought that could be used for a detailed observation. The analysis concentrates on the first two phases of CWA; Work Domain Analysis; and Control Task Analysis and on the fourth phase; Social Organisation and Cooperation Analysis. The analysis focuses on describing the way the system is currently configured, and at the same time identifies the freedom and the constraints within the system.

The chapter illustrates that at an early preproduction level the tool supports the CWA process. Particular encouragement was taken from the time taken to complete the process. The initial analysis was completed in one working day with the aid of the tool. Other key benefits identified include the dynamic nature and the ability to adjust the products rapidly. Valuable recommendations for further tool developments were recorded because of this analysis.

CWA of an Armoured Battlegroup in the Quick Attack

The following analysis is taken from a training video describing an Armoured Battlegroup in a Quick Attack scenario (BDFL 2001). The training video is non-specific to any particular regiment, takes place in a fictional setting, and can be roughly considered as conventional (symmetrical) attrition-based warfare. The scenario describes a battlegroup moving to engage OPFOR (opposition forces). On the approach to engage the main OPFOR outlined by intelligence, an OPFOR artillery cell is detected by a reconnaissance unit. At this point an order is given for the battlegroup to split down to engage the detected cell. This analysis focuses on the section of the battlegroup assigned to the OPFOR artillery cell. The faction assigned to this task includes:

A company consisting of:

- Two troops of Challenger II tanks (2 × 3 units): The tanks are one of the most versatile assets to the system; they can perform in; anti strong point, anti armour as well as securing open country. They also play a significant role in OPFOR detection.

- One Milan section (3 units): The main purpose of the Milan sections is in an anti armour role. They also have a role in OPFOR detection.
- One section of Armoured Engineer Field troops (4 × 432's): These units have the ability to shape the terrain by adding in and removing obstacles. They can also support in OPFOR detection.
- Half of the Recce Platoon (4 Scimitars): The main role of the reconnaissance unit is to detect and monitor the OPFOR, a significant part of this is also recording negative sightings to indicate where the OPFOR are not situated.

In addition, the following resources are also assigned from the battlegroup;

- HQ Warriors: The primary role of the HQ warriors is in command and control of the battlefield, they; receive orders from higher echelons, perform battle space area evaluation, identify strengths and weaknesses, and command subordinate units. The HQ warriors also play a role in OPFOR detection.
- Forward observation officer (FOO) in a Warrior OPV (Observation Post Vehicle). The FOO performs OPFOR detection and target acquisition. This information is used to direct fire.
- Mortar Fire Controller (MFC) in a Spartan performs a similar role as the FOO but only directs the mortars.
- Two platoons of four Warriors: The Warriors have the role of transporting the infantry. The infantry in turn have the role of securing close country as well as supporting in OPFOR detection.

Once the decision to engage has been made, the movie breaks the mission down into three distinct phases, these are:

- Find – determine the location and capabilities of the OPFOR
- Fix – prevent the OPFOR from moving, place them under pressure
- Strike – engage the OPFOR.

In addition to these phases, there is an initial phase before the OPFORs are detected and a consolidation phase after the strike has been completed.

Advance to Target

The role here of all units is to progress towards the target and to detect OPFOR.

Find

The Find phase starts as soon as the OPFORs are sighted. The activities taking place within the Find phase include:

- Identify OPFOR strengths and weaknesses: Information is used based upon location, combat effectiveness and capabilities.
- Weapons: An understanding is developed based upon the capabilities of weapons.
- Identification of Fields of fire: Fields of fire are identified based upon the range of munitions.
- Identify routes into OPFOR territory: Through the process of BAE, optimal routes into OPFOR are developed.
- Identify OPFOR withdrawal routes: Using the BAE products withdrawal routes are considered based upon terrain and doctrine.
- Identify Forming Up Position (FUP). The FUP is sought based on the following criteria: In dead ground, large enough to contain forces in assault formation, not subject to fire from flanks, as close to objective as possible without revealing intentions, as well as being a definite recognisable feature.
- Identify Fire Support Group (FSG). The FSG is responsible for relaying information on capability and availability ordnance and munitions. The FSG consists of one group of three Challenger II tanks, a FOO and the MFC.
- Detect Flank vulnerability. As part of the BAE any flank vulnerability is detected.
- Identify Flank protection areas. Areas are identified that require action to secure them from OPFOR threat.

Fix

Once the OPFOR have been surveyed, the next stage identified by the movie was the Fix phase. The activities taking place within the Fix phase include:

- Observation and fire: The OPFOR are engaged at long range while observation takes place.
- Lay-down fire from FSG (jockey and sniper OPFOR until H-Hour): The OPFOR are placed under pressure by continual long range fire.
- Issue warning orders: The aim of the warning order is to ensure that all units are aware of operation and their relevant roles.
- Secure FUP: The FUP is secured either by observation and fire in open ground or physical force in closed country.
- Flank Security: Flank security is put in place.
- Issue short concise confirmatory orders: The warning orders are confirmed shortly before moving. These orders consist of; confirmation of mission and tasks, location of OPFOR, location of FUP, location of FSG, location of flank security, other key locations, detailed timings, fire plan, and the H-Hour (the point where troops cross line of departure).

- Form up. The units get into battle positions before the transit to engage the OPFOR.

Strike

At H-Hour, the point where troops cross line of departure, the FUP the battle moves into the strike phase. The strike phase can be further broken down into three phases.

1. Prepare
2. Covering Fire (Approaching Target)
3. Defensive Fire (Engaging Target).

Strike Preparation

The FOO and Operational Commander (OC) identify OPFOR withdrawal routes and potential counter attacks. While forming up, the OPFOR are bombarded with salvos from the tanks, artillery units (AS90's) and mortars with the aim of causing as much damage as possible before contact. The OC and the FOO then join the assault group leaving the MFC at the FSG. The second troop of tanks (those not at the FSG) 'leads' the assaults (the intimate support troop). Two infantry assault platoons form up behind. The company HQ units form up in the centre. The third infantry platoon commanded by the 'Second in command' (2IC) acts as reserve.

Covering Fire

The FOO calculates how long covering fire will take place after H-Hour. The aim is to deliver infantry onto OPFOR as swiftly as possible, in the correct formation, with as few casualties as possible. The intimate support tanks 'lead' the infantry concentrating fire on the OPFOR positions. If the attack halts for more than two minutes, the infantry will need to dismount for protection and observation.

Defensive Fire

As the strike force approaches, the fire support switches to depth targets. The assault group begins through clearance, infantry dismount at forward edge. The Warriors either:

- Move to rear – providing covering fire
- Move to flanks – providing flank security or fire onto objective
- Move to dead ground – awaiting orders
- Stay and fight through.

Fight through should be thorough and controlled. Momentum is maintained to prevent the OPFOR from gaining initiative, committing the reserve is often the key to success.

Consolidation

The aim of the last phase is to secure the physical perimeter and to consolidate the current position (evacuate POW, Casualty Evacuation, Vehicle Evacuation)

Secure

- Hasty defensive positions need to be secured
- Fire support tanks move forward to join intimate support tanks, they aim to move beyond objective to produce ring of steel
- Warriors and infantry with 94mm LAW take up defensive positions.

Consolidation

- Battlefield clearance
- Engineers – clear and destroy key OPFOR equipment
- Engineers – Prepare defensive positions for infantry
- REME – vehicle services and vehicle evacuation
- The Command Sergeant Major (CSM) sets up a company Rendezvous (RV) at the forward edge of the objective to collect casualties, POW, Combat supplies.

The Analysis

The Abstraction Hierarchy

The first phase of CWA, the Abstraction Hierarchy, is used to describe the domain in event independent terms. Here the physical units and their capabilities are captured. At the top of the hierarchy, the overall purpose of the system is recorded and decomposed. The procedure for applying Naikar et al's (2005) nine-step methodology for completing an AH was as follows:

Establish the purpose of the analysis   The purpose of the analysis is to validate and identify further requirements for the CWA software tool. The purpose is also to describe the scenario in enough detail that it can be used to showcase the CWA framework.

Identify the project constraints   The analysis is constrained by the information on the video. To maintain credibility a conscious decision was made to base the analysis solely on an existing accepted document (BDFL, 2001).

Identify the boundaries of the analysis   To test the efficiency of the tool the analysis was limited to a single day, an analyst team was constructed with three analysts; one acting as a CWA expert, one as the CWA software tool expert and the third as a domain expert.

Identify the nature of the constraints in the work domain   The analysis is based solely on the described units, the analysis starts at the point these break away from the main group and ends at the end of the strike phase.

Identify the sources of information for the analysis   All information was extracted from the training video (BDFL 2001). The domain SME was on hand to give advice on explaining acronyms and unit capabilities although great care was taken not to extrapolate data beyond the contents of the video.

Construct the AH with readily available sources of information   In this case the content for Abstraction Hierarchy was extracted from the content of the training video. The video, lasting approximately 20 minutes, was viewed a number of times. The first stage of the analysis was to identify the functional purpose of the system. The introduction to the video made this clear, although minor alterations were made to semantics during the analysis, the final functional purpose was determined to be; 'To Efficiently Eliminate OPFOR Combat Effectiveness'. Next, following the structure of the video, each of the assets were introduced – these forming the majority of the physical objects in the AH. As each of the units was introduced, a basic description of their capabilities was given – this formed the basis for the object related processes. The 'values and priority measure's and the 'purpose related functions' were less easy to directly extract from the video. The video had to be viewed a number of times to capture all of the required information.

Construct the AH by conducting special data collection exercises.   An initial version of the AH was completed based upon the numerous viewing of the video. As specified in the project constraints, it was decided not to extend the analysis beyond the data contained in the video.

Review the AH with domain experts.   The analysis was validated in discussion with the domain SME. Additional capabilities were added to the physical objects that were less explicitly described in the video. The semantics of the values and priority measures were also corrected as well as some of the nodes being joined, whereas, others were subdivided.

The final value priorities and measures were determined to be: Simplicity; Maintain accurate situation awareness; Speed of execution (Tempo); Acquire terrain; Minimising casualties; and Maximum concentration of combat power;

The final purpose related functions describe the object related processes in relation to the overall purpose of the system. These were determined to be: Mission analysis; Prepare battle plan; Build and maintain picture of OPFOR situation; Coordinate resources; Provide covering fire; Defend position; Eliminate OPFOR armour combat effectiveness; Eliminate OPFOR strong points combat effectiveness; Eliminate OPFOR Infantry combat effectiveness Improve terrain; Remove POWs; Human CAS EVAC; and Vehicle CAS EVAC

Validate the AH   The AH was validated by reconfirming each of the means-ends links. Each node was taken in turn and the current links validated with the SME, each of the other possible links was explored to confirm a correct rejection. For thoroughness, this process was completed once and then revalidated after a short break. The final version of the Abstraction Hierarchy is presented in Figure 9.1. An example of the why-what-how triad can be seen by taking the node 'fire direction' in the object related processes layer to addresses the question of 'what' needs to be achieved. By tracing the links above, it can be seen that fire direction is required (the why) to fix the OPFOR, provide covering fire, defend the position, and eliminate OPFOR combat effectiveness. Tracing down it, the 'how' can be seen to include the FOO and the MFC.

The Abstraction Decomposition Space

Each of the nodes in the AH can be categorised into the different levels that the system is decomposed to, in this case; total system; sub-system; or component. This adds an additional dimension to the representation known as the Abstraction Decomposition Space (ADS).

The ADS does not reproduce well in A4 format in its current state, this has been identified as an area for further development.

The Contextual Activity Template

The system can be further described using the Contextual Activity Template (see Figure 9.3). The Contextual Activity Template is a two dimensional space that plots situations within the mission against functions that the system is performing. In this case, the situations have been taken from clearly delineated sections of the process defined by either time location or a composite of the two. A situation can be broadly defined as a point where there are significant role changes for the components within the system. The functions of the system have been taken from the Purpose related function level of the AH.

The dotted boxes indicate the range of situations in which the functions can effectively occur. The solid line indicates the situations where they typically

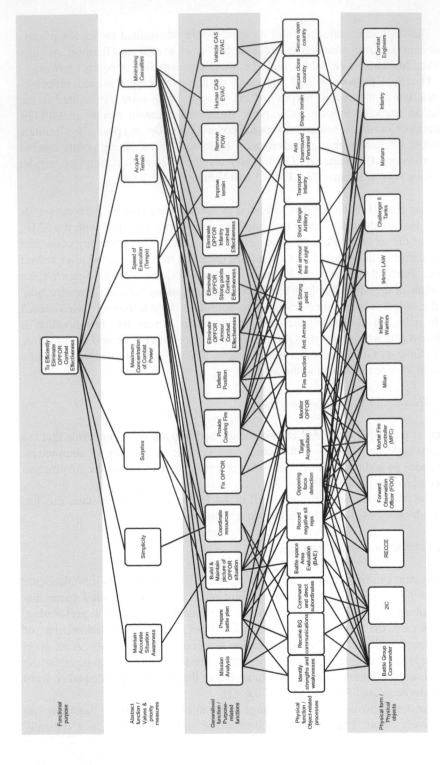

Figure 9.1 Abstraction Hierarchy for a battlegroup in a quick attack

occur. The difference between these two represents the adaptive boundary of the system. In effect the inherent flexibility. The 'Vehicle evacuation' function at the bottom of the table can occur anytime; however, it will typically take place in the consolidation phase of the mission. The function 'Coordinate resources' is a consistent activity; the template shows this by indicating that it typically takes place in all situations. Other activities can be seen to stop as another occurs; in this case the OPFOR is fixed up until approaching the target. At this point, it is no longer safe to fix the OPFOR and the activity 'provide covering fire' commences. Based on this analysis, in this scenario the system is only using 59 per cent of its adaptiveness (see Figure 9.4).

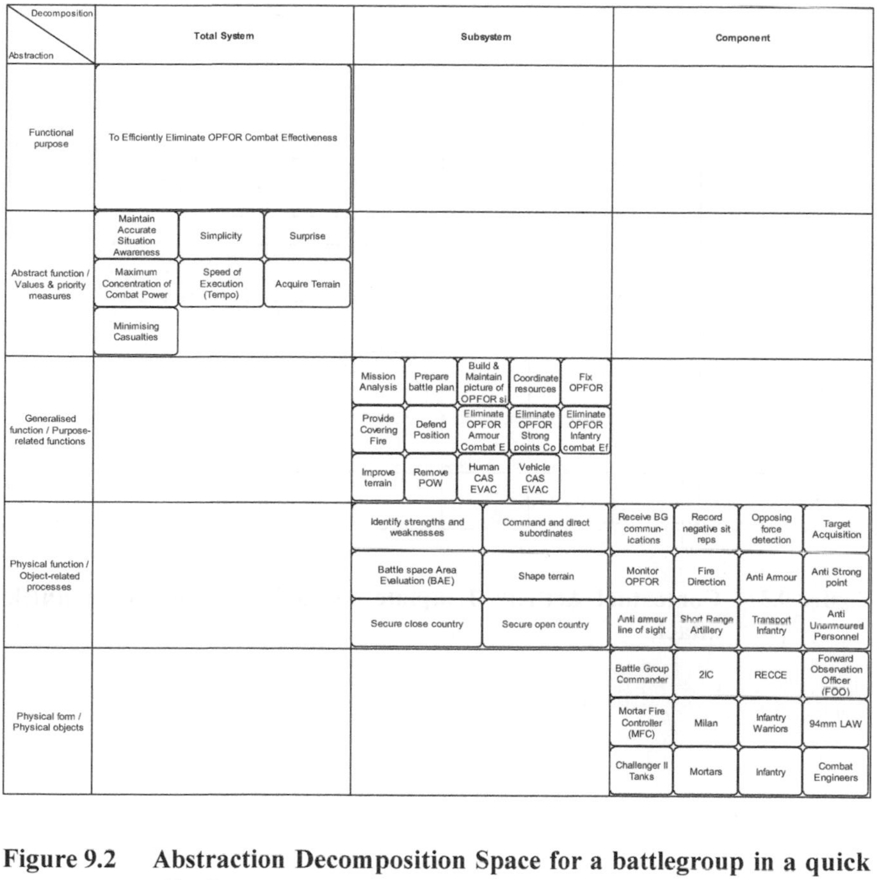

Decomposition / Abstraction	Total System			Subsystem			Component			
Functional purpose	To Efficiently Eliminate OPFOR Combat Effectiveness									
Abstract function / Values & priority measures	Maintain Accurate Situation Awareness	Simplicity	Surprise							
	Maximum Concentration of Combat Power	Speed of Execution (Tempo)	Acquire Terrain							
	Minimising Casualties									
Generalised function / Purpose-related functions				Mission Analysis	Prepare battle plan	Build & Maintain picture of OPFOR si	Coordinate resources	Fix OPFOR		
				Provide Covering Fire	Defend Position	Eliminate OPFOR Armour Combat E	Eliminate OPFOR Strong points Co	Eliminate OPFOR Infantry combat Ef		
				Improve terrain	Remove POW	Human CAS EVAC	Vehicle CAS EVAC			
Physical function / Object-related processes				Identify strengths and weaknesses		Command and direct subordinates		Receive BG commun-ications	Record negative sit reps	Opposing force detection	Target Acquisition
				Battle space Area Evaluation (BAE)		Shape terrain		Monitor OPFOR	Fire Direction	Anti Armour	Anti Strong point
				Secure close country		Secure open country		Anti armour line of sight	Short Range Artillery	Transport Infantry	Anti Unarmoured Personnel
Physical form / Physical objects								Battle Group Commander	2IC	RECCE	Forward Observation Officer (FOO)
								Mortar Fire Controller (MFC)	Milan	Infantry Warriors	94mm LAW
								Challenger II Tanks	Mortars	Infantry	Combat Engineers

Figure 9.2   Abstraction Decomposition Space for a battlegroup in a quick attack

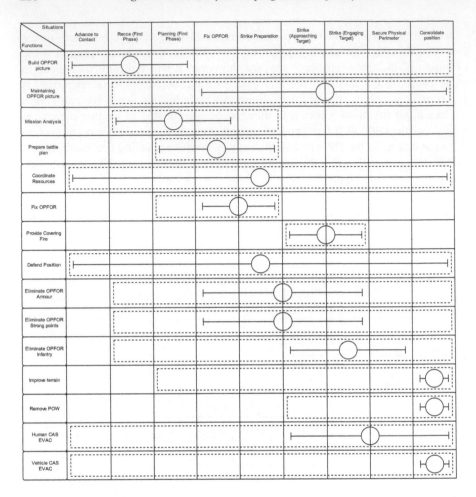

Figure 9.3   Contextual Activity Template for a battlegroup in a quick attack

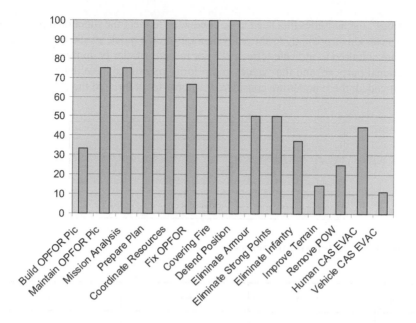

Figure 9.4 **Contextual activity breakdown showing where functions did occur compared to where they 'could' occur. 100 per cent indicates the adaptive boundary, the bars present the percentage used for each function**

Social Organisation and Cooperation Analysis

Social Organisation and Cooperation Analysis takes the existing products generated; the Abstraction Decomposition Space, and the Contextual Activity Template and codes them to indicate the constraint limits of each of the actor groups.

In this example the actors within the system have been classified into one or more groups; sensors, effecters and coordinators (see Figure 9.5 for key)

Sensors are units that are used to detect targets (such as reconnaissance units); effecters are units that can have an effect on the OPFOR or on the terrain (such as artillery or engineers). Coordinators, as the name suggests, have a primary role of coordinating forces (such as Commanding officers, Mortar fire controllers). This categorisation requires the analysis to carefully consider each actor, it could be argued that at some level, each of the units does each of the roles; however, categorising the units in this way would defeat the objective of categorisation.

In the 'physical objects' level the Abstraction Decomposition Space coloured in Figure 9.6 indicates which actors are considered to be sensors, effecters, and coordinators. The diagram also attributes the 'object related processes' performed to one or more of these groups. Finally the diagram also shows the 'purpose related functions'; the activities that need to be conducted to meet the functional purpose of the system. Unsurprisingly as the analysis focuses on a quick attack, the majority of cells are coloured to indicate 'effecters'. For reasons of clarity, in this example, the 'functional purpose' and the 'values and priority measures' have deliberately been left un-coded; these nodes are considered to apply to all three of the groups

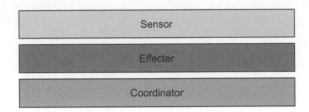

Figure 9.5 Actor group colours

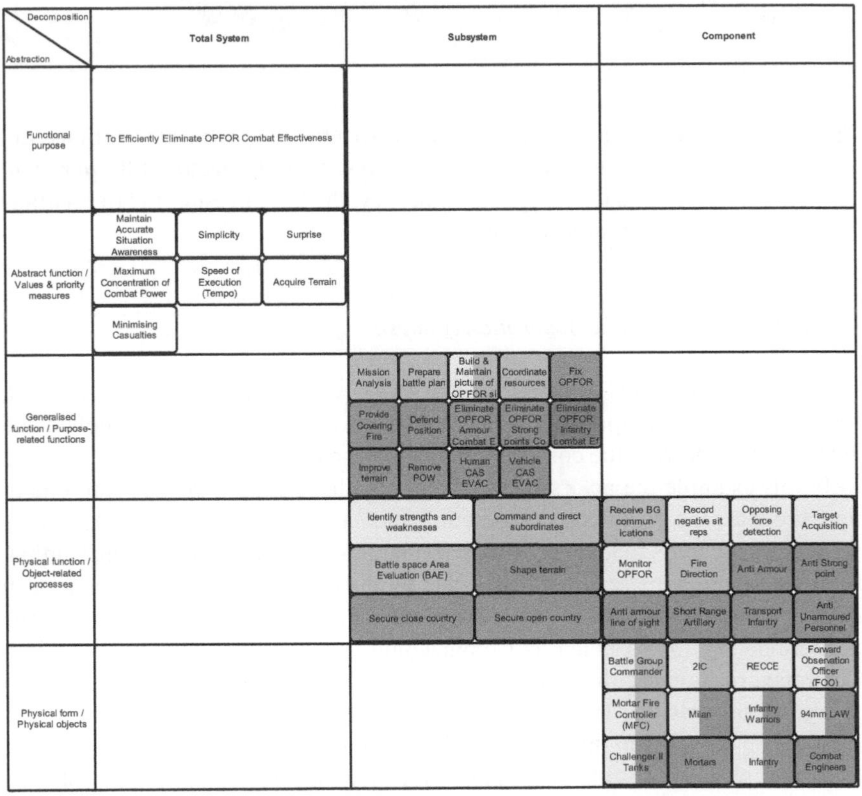

Abstraction \ Decomposition	Total System	Subsystem	Component
Functional purpose	To Efficiently Eliminate OPFOR Combat Effectiveness		
Abstract function / Values & priority measures	Maintain Accurate Situation Awareness; Simplicity; Surprise; Maximum Concentration of Combat Power; Speed of Execution (Tempo); Acquire Terrain; Minimising Casualties		
Generalised function / Purpose-related functions		Mission Analysis; Prepare battle plan; Build & Maintain picture of OPFOR sit; Coordinate resources; Fix OPFOR; Provide Covering Fire; Defend Position; Eliminate OPFOR Armour Combat E; Eliminate OPFOR Strong points Co; Eliminate OPFOR Infantry combat Ef; Improve terrain; Remove POW; Human CAS EVAC; Vehicle CAS EVAC	
Physical function / Object-related processes		Identify strengths and weaknesses; Command and direct subordinates; Battle space Area Evaluation (BAE); Shape terrain; Secure close country; Secure open country	Receive BG communications; Record negative sit reps; Opposing force detection; Target Acquisition; Monitor OPFOR; Fire Direction; Anti Armour; Anti Strong point; Anti armour line of sight; Short Range Artillery; Transport Infantry; Anti Unarmoured Personnel
Physical form / Physical objects			Battle Group Commander; 2IC; RECCE; Forward Observation Officer (FOO); Mortar Fire Controller (MFC); Milan; Infantry Warriors; 94mm LAW; Challenger II Tanks; Mortars; Infantry; Combat Engineers

Figure 9.6 Abstraction Decomposition Space coloured by actor type

The Contextual Activity Template can be coloured in the same way. This representation focuses on activity, whereas, the Abstraction Decomposition Space focused on the functions within the system. The focus on activity further illustrates the dominance of the effecters within the system in the quick attack context.

Conclusions

This chapter has documented an exercise intended to validate and extract key requirements for the CWA software tool. The chapter also stands as a demonstrator of the framework in the first two phases (WDA, ConTA) and the fourth phases (SOCA) of CWA.

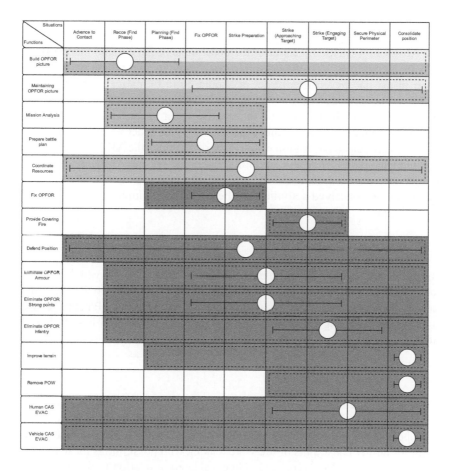

Figure 9.7   Contextual Activity Template coloured by actor type

Methodological Issues and Findings

The process of following the CWA, allowed the analysts to gain a large degree of domain understanding in a very limited period. The focus on constraints prevented the analysts from becoming over concerned with low-level detail, often referred to as 'going down too many rabbit holes'. The final products of this analysis serve as a concise, constraint-based description of the system. The process of modelling the system constraints is very important; it leads the analysts to question, why the constraints are in place and how the system would be different if these constraints were to be removed. The analysis provides a good model to determine the effects of manipulating constraints (adding, removing or changing) on the wider system. The model captures in broad terms the redundancy within the system and the potential for informing reallocation decisions.

AH/ADS

The Abstraction Hierarchy captures the constraints of the domain independent of activity, it addresses what the complete system and its constituent parts could do rather than what they currently do, or should do (a formative description rather than a normative or descriptive description). The model also considers the system in terms of its overall objectives; it focuses the reader to consider how any activity within the domain influences the system's ability to enact its overall functional purpose. Through consideration of the means-ends links the AH provides an arena to explore the redundancy and flexibility within the system. As the model is event independent and indeed unconstrained by the infrastructure, new unplanned, previously un-conceived activities can be considered within this existing model of the domain.

CAT

The CAT captured the constraints and freedoms of the way in which activity is conducted. The difference between, where activity 'is conducted' and 'can be conducted' illustrates the freedom and adaptability of the system. Potential changes to Standard Operating Procedures (SOPs) could be considered based upon this model.

SOCA

The SOCA phase classifies each of the actors in the system into one or more of the following categories; sensor, effecter; commander. The analysis goes on to describe activity by the type of actor needed to conduct it. This stage of the analysis identifies further constraints of the system along with possibilities for reallocation of resources. The actors' types are also considered in relation to the phase of the battle, introducing a temporal and situational element to the analysis.

The model developed describes a system that is tightly controlled by SOPs and protocol. The levels of flexibility and adaptability within this military domain are considered less than in many civilian counterparts. The AH and the CAT models provide a product with which to explore the potential for flexibility and task reallocation.

This chapter provides further validation for CWA as a comprehensive framework of tools. The majority of previous CWA applications have utilised only the initial WDA phase. We contend that future CWA should utilise the phase that is most applicable to both the research aims and the domain in question. This analysis demonstrates the interconnection between the CWA phases and reinforces the benefits of extending a CWA beyond the initial Work Domain Analysis and Control Task Analysis.

Software Recommendations

The process was successful in its main aim; the software tool was able to support the development of the required documentation and some clear recommendations for incremental changes were made for the tool improvement.

Encouragement was taken from the software tool's ability to allow a comprehensive analysis to be conducted by a small analyst team is a single working day. It should be noted; however, that the process was significantly expedited by the fidelity of the data. Much of the analysis time appears to be in refining the data into a usable format. The high-level training video was ideal for this purpose as it presented data at an appropriate level of fidelity for this analysis.

Future work is underway implementing the recommendations for the tool elicited from this process and further similar validation studies form part of the wider validation phase of the software tool's development.

Chapter Summary

This chapter has evaluated the software tool against a real military scenario. The tool performed well and showed many clear advantages over both the paper-based alternative as well as alternatives in other software tools such as Microsoft Visio. The tool supports the iterative process of documentation well. At the start of analysis, it is very difficult to predict the size or layout of the representations making it very difficult to create them on paper. The tool also has clear advantages over drafting software packages such as Microsoft Visio; the ability to pass information forward within the phases was extremely valuable in the process of expediting the documentation. The functions for the CAT were imported directly from the AH; the AH and the CAT products were reused in the SOCA phase. The ability to feed-forward semantic changes to related diagrams also proved to save significant amounts of time. Possibly the greatest advantage of the tool over 'MS Visio' was discovered in the validation phase of the Abstraction Hierarchy. During the validation process, each node was considered in turn. By 'mousing over' the node, the tool highlights the related nodes in the levels above and below in red. This had a significant impact on the time taken and the ease of validating the links. The time taken to generate the products and the ease of manipulation of the diagrams meant that the use of the tool did not act as a significant barrier in the creative process of conducting the analysis. Diagrams could be rapidly rearranged to improve their clarity, with the changes to layout reflected in later phases. The ability of the tool to support the training of novices was not explicitly tested in this analysis; however, it is perceived that this would further justify the development of the CWA software tool.

Chapter 10
Conclusions

The current literature has highlighted that for such dynamic unpredictable environments, many existing normative modelling approaches have their limitations. Much of the body of literature studied, proposes CWA as an appropriate framework for the analysis, development and evaluation of these complex domains.

As stated in the introduction, this book builds heavily upon the previous, pioneering, work on CWA (most notably; Rasmussen et al, 1994; Vicente, 1999). A concise effort has been made to take these descriptions of the framework, develop them, and describe them in detail that is more explicit. The book has taken what was a non-prescriptive framework, and, through the use of case studies, provided a clearer description of a process for conducting an analysis. In an attempt to communicate the benefits of the CWA approach, the framework has been described in both familiar and complex domains; the analysis of these domains has highlighted the capability of the approach in; the extraction of design recommendations and system evaluation. The benefits of applying the CWA framework in its entirety have also been explored and advancements have been made in the explicitness of the links between these phases. Using case studies, an approach detailing the process of extracting design recommendations from an analysis has been presented. A new, theoretically grounded, approach for the evaluation of complex sociotechnical systems is also introduced.

The common concern with the CWA approach, 'that it cannot inform design', has been challenged. The importance of a structured approach to support the creative process of the design for complex systems is discussed. The fundamental principles of interaction design have been introduced and applied to the development of a number of interfaces. This link between analysis and design has then been explored for the development of rapidly reconfigurable interfaces to support a 'sensor to effecter' paradigm and the dynamic allocation of function. The approach has also been used to inform the development of a number of military decision support tools, with the aim of supporting rough order of magnitude planning.

With a non-prescriptive approach, questions will always be raised about the ability to teach and learn the technique. It is contended, that with the detailed description of the individual phases, and the clarification of the representation products used, this book enhances the assimilation of the approach. The more descriptive approach, along with the selected representations, has been encapsulated in a bespoke software support tool; assisting novices and expediting the documentation process. The application of the developed software tool has been validated against a complex military scenario.

Through the production of this book, a number of key questions were generated; these questions form the basis of the summary of the work discussed:

Is CWA Useful for the Analysis, Development and Evaluation of 'Command and Control' Systems?

The initial chapters to this book introduced the concept of complexity; this notion of complexity is then discussed within the domain of command and control. These initial chapters raise concerns relating to normative modelling of such complex dynamic domains. CWA is proposed as a framework for coping with this level of complexity. The approach has been applied to a number of case studies throughout this book; each of these case studies revealed positive benefits of the CWA framework. Chapter 5 uses two case studies to explore each of these issues in turn. The first analysis, based upon a helicopter mission-planning tool, provides an example of how design recommendations can be readily extracted from a comprehensive analysis of system constraints. In the second case study, investigating military battlefield planning, a new approach to evaluation is presented. Benefits explored throughout this book include: the exploration of constraints (applied throughout); the modelling and manipulation of system flexibility (applied throughout; specifically, see iPod example in Chapter 4); the allocation of function (see Chapter 6); the extraction of design recommendations (see Chapters 4 and 5); and system comparison and evaluation (see Chapter 5).

In summary, this book has shown a number of clear uses of the CWA framework in the roles of analysis, development and evaluation. Based upon the description of the CWA framework, presented in Chapter 2, a number of clear benefits of the approach, over other normative approaches, are extracted. It is contended that developments to the framework, proposed in this book, have enhanced the use of the tool and made the current benefits more explicit.

Can it Cope with Complexity and Different Units of Analysis (Artefacts to Dynamic Systems)?

The application of CWA was first introduced to this book, in Chapter 4, at an artefact level. In Chapter 5, more complex, 'systems of artefact' domains were modelled. In Chapters 6 and 9, larger 'systems of systems' were then studied. Whilst the approach performed well in each, it is fair to say that the benefit of the approach at an artefact level is limited; it is at the systems level that the power of the approach is revealed. Further evidence to support this claim can be found at an artefact level (e.g. Crone, 2003; Jenkins et al, 2007d); at a systems of artefacts level (e.g. Ahlstrom, 2005; Hadjukiewicz, 1998; Jamieson and Vicente, 2001; Jenkins et al, in press a; and Miller, 2004); and at a systems of systems level (e.g. Burns et al, 2000; Hadjukiewicz et al, 1999; Jenkins et al, 2007c and Lintern, 2006). This book has shown that the benefit of the tool is related to the complexity of the system. With simple artefacts, such as the iPod, the benefit of the approach can be seen to be limited; however, in complex environments, with multiple constraint sets, such as the mission-planning tool (Chapter 5), the battlefield management system (see Chapter 5) or the sensor to effecter paradigm (see Chapter 6) a constraint-based approach, such as CWA is, almost essential.

What Are the Limits of CWA?

Like the domains it was developed to model, CWA, as an approach, can be described as complex. To fully understand the framework the user is required to depart from goal and task orientated descriptions of work, that they, are frequently, more comfortable with. Much of the skill in applying CWA lies in setting the initial boundaries for the analysis, if the boundary is incorrect, then the analysis produced is likely to be incomplete, inconclusive or misleading (see Hajdukiewicz et al, 1999). As the framework continues to be applied to more domains and used for different applications, the body of literature on CWA grows. What was once an approach that was only accessible to disciples of Jens Rasmussen or Kim Vicente, is now fathomable by anyone willing to devote the time and patience required to the study of the framework.

As discussed throughout the book, the framework has its limitations. It is not intended to be used in isolation. The synergies that exist between CWA and other more established human factor methods, has been briefly explored in this book (Chapter 2). The analysis of the MPS system has explored the compatibility of CWA with approaches such as, EAST (Walker et al 2006a), Social Network Analysis (Houghton et al, 2006), and Distributed Working (Stanton et al, 2006). In all cases, CWA provides a less descriptive, event-independent approach. It was generally found that this was complimented by the more detailed normative approaches just discussed. A particular benefit was identified in the combined application of SNA and CWA. SNA is very good at describing links between agents in an environment, mainly through their frequency – from the application of network statistics it is possible to identify key agents; the limitation of the SNA technique; however, is its failure to identify redundancy within the system. The SNA approach only models links that have been observed, rather than the links that could possibly exits. It is contended that the synthesis of this approach with CWA (specifically WDA) could provide a validation of the importance of links. By examining the redundancy within the system, it would be possible to identify which links are truly pertinent and therefore, 'key'.

Developments and Directions for Future Work

With a non-prescriptive framework, it is important to develop a library of case studies; each new application of the approach provides additional insight into the capabilities and limitations of the approach. Of particular interest are applications that move beyond analysis, and attempt to use CWA to inform and lead design or evaluation. As stated in this book, when compared to the later phases, the phases of WDA and ConTA have received unequal attention in the current literature. Further work is required in applying these later phases to new domains, exploring their additional benefits.

Work Domain Analysis (WDA)

The first phase, Work Domain Analysis, has been applied in the widest number of domains and for the widest number of applications, because of this; it is fairly well understood with consensus on its application. Whilst much is clearly understood about this phase of the analysis, this book has demonstrated that the potential for further exploitation exists. This book has presented clear advancements in the use of this phase as a template for rough order of magnitude planning (see Chapter 7). Arguably, through the realisation that the upper levels of the abstraction hierarchy present technologically agnostic description of the systems, a more notable and transferable advancement has been made. An approach has been developed based upon the abstraction hierarchy for complex system comparison and evaluation (see Chapter 5). Additional, novel, benefits highlighted in this book include the use of the abstraction hierarchy for data clustering in interface design and the development of training plans based upon different levels of abstraction (see Chapter 5). Another clear advancement is the direct extraction of interface requirements from the nodes in the abstraction hierarchy (see Chapter 6).

The dynamic interface developed within the CWA software tool has significant benefits for this phase; the ability to rapidly generate and re-order abstraction hierarchies allows representations to be generated during real-time discussions with SMEs. The dynamic nature of the tool, and its ability to highlight linked nodes, also makes the job of validating the means-ends links much easier.

Control Task Analysis (ConTA)

The second phase, Control Task Analysis (ConTA), has received, comparably, much less attention than the WDA phase, but more than the remaining phases. Its use of the widely accepted, and understood, decision-ladder has made its adoption possible, with ever-increasing interest in this phase. The work of Naikar (in press) on the Contextual Activity Template (CAT) has been embraced in this book, this graphical representation has been adopted and applied to a number of new situations providing important validation for the development of the phase.

Notable advancements made in this book include the exploration of flexibility and adaptiveness within the CAT (see Chapters 3, the analysis of an iPod and Chapter 9, the analysis of a Battlegroup in a quick attack), and more specifically the extraction of product modifications required to enhance flexibility (see the analysis of an iPod and Chapter 9). Another notable advancement is the use of the decision-ladder to extract directly, contextually applicable, interface information requirements (see Chapter 6)

Strategies Analysis (StrA)

The third, and perhaps least well defined phase; Strategies Analysis (StrA), is possibly the most in need of further consideration. In this book, a conscious decision has been made to move away from the extremely context-dependant flow maps proposed by Vicente (1999); the more structured representations proposed by Ahlmstrom (2005) are favoured. Whilst this book has not attempted to develop this phase of the approach significantly, notable advancements include the adoption and development of Ahlstrom's (2005) representation. Additional developments include the use of the decision-ladder to explore in detail each strategy (see Chapter 3). Finally, an advancement is made in the use of the CAT to determine the strategies requiring further analysis (see Chapter 3, the analysis of an iPod).

Social Organisation and Cooperation (SOCA)

The Social Organisation and Cooperation (SOCA) phase of the framework is, most probably, the phase that has been advanced the most through the writing of this book. The modelling of the actors onto the previous three phases is nothing new; in fact, Vicente (1999) advocates this approach. What is new; however, is the approach for doing this. Vicente (1999) proposes a much more general, less detailed approach. This book has seen the benefits of combining a common coding approach with new representations such as the CAT. This approach describes the constraints that are instrumental in the allocation of activity for particular scenarios (this is expressed most clearly in the MPS example in Chapter 5). As with any new approach, this representation approach would benefit from further application and validation. Another clear advancement in this phase is the use of the decision ladder, at a system level, as a model for the dynamic allocation of function (see Chapter 6).

Worker Competencies Analysis (WCA)

Finally, the fifth phase, Worker Competencies Analysis (WCA), has also received little attention from the literature. Whilst the theory behind the SRK taxonomy is widely understood, its application as part of the CWA framework is much less so. Kilgore and St-Cyr (2006) make a very welcome advancement in the representation of this phase, presented in their conference paper. However, at the time of publication this approach lacked applications and validation. This book has built on the work by Kilgore and St-Cyr (2006); extending their table so that it includes a graphical link to the process steps modelled in the decision-ladder (see Figure 2 13). The description in this book also describes a much closer relationship to the decision ladders generated in the ConTA phase. It is noted, however, that this approach still requires significant additional validation.

Linking the Phases

Arguably, one of the most visible advancements made by this book is the development of the explicit links between the five phases of analysis. A strong emphasis has been placed upon the need to extend analyses beyond the initial phases in order to capture the constraints described by the entire framework. In the analysis of complex sociotechnical domains, the study of the constraints governing interaction and the allocation of function is essential. It is perceived that the more explicit approach, described in the CWA software tool, used to link the phases will encourage analysts to continue their analysis into the later phases. The following list introduced in Chapter 8 details the explicit links designed into the CWA software tool.

1. The nodes created in the AH are copied through to the ADS where they are allocated to a level of decomposition, any subsequent changes to the AH carry forward.
2. The option exists, within the software tool, to import the text from the nodes in the AH to the CAT. This allows the link between the physical functions level of the abstraction hierarchy and the functions within the CAT to be exploited. There is deliberately no dynamic link for this so subsequent changes to the AH are not reflected in the CAT.
3. The tool creates a mini decision-ladder for each cell in the matrix in which activity can take place.
4. Information for each activity can then be described in detail in a larger decision-ladder.
5. For each stage in the strategies analysis there exists, the option to create a decision-ladder, the software automatically labels the decision-ladder and places a diagram at the top showing the relevant strategy.
6. The ADS is used in the SOCA phase for coding, the link between the figures is constant so subsequent changes are reflected in the SOCA phase.
7. The CATs and decision-ladders are used in the SOCA phase for coding, the link between the figures is constant so subsequent changes are reflected in the SOCA phase.

8. The strategies analysis flow diagram is used in the SOCA phase for coding, the link between the figures is constant so subsequent changes are reflected in the SOCA phase.

9. Information from the notes made on the decision-ladders can be automatically imported to the WCA matrix.

Whilst not included in the current version of the CWA software tool, the direct exploration of the functions in the CAT through strategies analysis, and the subsequent representation in the CAT is also considered an important link established in this book.

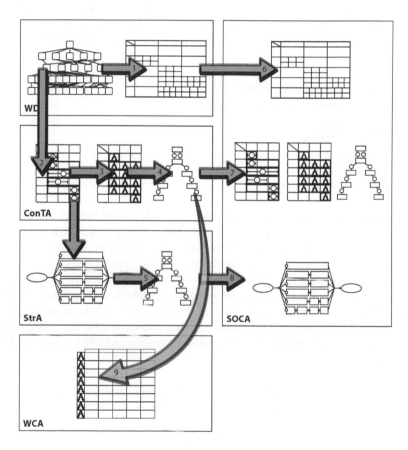

Figure 10.1   Links between phases supported by the CWA tool

Extending CWA

The success of any analysis technique can be linked to its ability to solve real world problems. By using an analysis output a number of times it is possible to amortise its development time and, more importantly, cost. Kirwan and Ainsworth (1992) claim that HTA can 'be used in almost every circumstance' (p. 29); Ainsworth and Marshall (1988) describe HTA as being 'perhaps the nearest thing to a universal task analysis technique'. Staples (1993) describes how HTA was used as the basis for virtually all of the ergonomic studies in the design and development of a nuclear reactor. Specifically, he details that the HTA approach was used for; identifying potential errors, interface design verification, identification of training procedures, development and verification of operating procedures, workload assessments, and communications analysis. Stanton (2006), in his paper on the HTA approach, claims at least twelve additional applications for HTA. Stanton (2006) describes seven of these extensions:

- Investigate design decisions
- Analyse human machine interaction
- Predict error
- Allocate function
- Design jobs
- Analyse team work
- Assess interface design.

Whilst the importance and prevalence of HTA within the Human Factors literature cannot be denied, Chapter 2 has shown that the approach does have its limitations. HTA is essentially a normative description of activity. It focuses on how activity is, or should be conducted. Unlike approaches such as CWA, HTA is not well suited for the analysis of first-of-a-kind systems. It is arguably less well suited to the modelling of complex sociotechnical systems (Jenkins et al, 2008a; Naikar and Lintern, 2002). Extensions of CWA have already been applied successfully to real world domains. Considering the list described earlier by Stanton (2006) a literature review reveals applications of CWA to each of these stages of the design process.

- Investigate design decisions (Cummings and Guerlain, 2003)
- Analysis human machine interaction (Jenkins et al, 2008a)
- Predict error (Naikar and Saunders, 2003)
- Allocate function (Jenkins et al, 2008a)
- Design jobs (Lintern and Naikar, 2000)
- Analyse team work (Naikar and Pearce, 2003; Lintern and Naikar, 2000)
- Assess interface design (Vicente, 1999; Burns and Hajdukiewicz, 2004)

In her paper, title 'beyond interface design', Neelam Naikar (2006) lists further extensions of CWA as the following:

- Training needs analysis and training-systems requirements
- Evaluation of system design proposals
- Team design
- Training strategies for managing human error.

Using the design process as a basis, Bonaceto and Burns (2003) explore how different methods can assist the different stages of design. This model has been used as a basis to generate Table 10.1; this table provide a framework for exploring future applications of CWA. The table lists each stage of the design process, explaining how CWA could be of use. The central column indicates the phases of analysis of primary importance. A short description is provided in the right-hand column. The table shows that CWA can be used at each stage of the process to assist design. Further exploration of some of the developments described is required.

Table 10.1   Applicability of CWA phases to stages of the design

Concept Definition • Determine the systems overall purpose • Identify system boundaries • Identify interactions of the system with its environment.	**WDA** ConTA StrA SOCA WCA	The WDA phase of CWA can help answer questions about why a new system should exist, what functions it should implement, and what physical devices are necessary. The analysis is technologically agnostic supporting new system design.
Requirements Analysis • Develop system requirements and specifications.	**WDA** ConTA StrA SOCA WCA	By applying metrics to each of the nodes in the WDA, requirements can be generated, the means-ends links can be used to trace how changes to these metrics is likely to influence other parts of the system.
Function Analysis • Define functions that will meet the system requirements.	**WDA** **ConTA** **StrA** SOCA WCA	The WDA phase of CWA can be used to represent the functional decomposition of the system. As this book has shown the purpose related functions level can be explored in greater detail in the Control Task Analysis phase, using the CAT, and the strategies analysis phase

Table 10.1   *Continued*

Function Allocation		
• Effectively distribute functions of the system between people and technology.	**WDA** **ConTA** **StrA** **SOCA** WCA	The SOCA phase of CWA can be used to describe actor roles, including the allocation of function and coordination structures. This phase is built upon the results from the ConTA, and StrA phases of CWA. These are grounded with the WDA, which contains context for the functions. This approach has been demonstrated in Chapter 6.
Task Design		
• Given constraints of the system's functional architecture, define how humans in system will carry out the tasks that have been assigned to them.	WDA **ConTA** **StrA** **SOCA** WCA	The ConTA phase describes the tasks, or functions, required, within the CAT; these are then explored in detail in the StrA. Finally, the allocation of resources is considered in the SOCA phase. This stepped approach allows the different constraint sets to be considered separately; providing the analyst with a more structured approach.
Interface and Team Development		
• Identify and develop designs and concepts for the interfaces between people, software, and other people.	WDA ConTA StrA **SOCA** WCA	The SOCA phase can be used to describe actor roles, including the allocation of function and coordination structures. Chapter 6 has shown how interfaces can be directly developed based upon the CWA.
Performance, Workload and Training Estimation		
• Evaluate physical and cognitive workload levels of individuals and teams with a proposed system design.	**WDA** **ConTA** **StrA** SOCA WCA	The WDA phase of CWA can be used to define the metric requirements of training simulators. A greater understanding of the tasks and the situations they are conducted within is contained within the ConTA and StrA phase.

Table 10.1 *Continued*

Requirements Review • Throughout the system development process, review the system design with respect to its requirements and operational need.	WDA ConTA StrA SOCA WCA	The WDA phase of CWA can be used to evaluate alternative designs in terms of how well the technical solution supports the functional purposes of the work domain. The evaluation technique proposed in chapter 5 is ideal for this purpose. CWA also allows testing of whether the system under development supports the necessary control tasks, strategies, role allocation and coordination structures, and operator's cognitive abilities.
Personnel Selection • Specify characteristics of personnel needed to achieve their tasks.	WDA ConTA StrA SOCA **WCA**	The WCA phase addresses the level of skill required by the human operator dependant on the system constraints and configuration.
Training Development • Develop training materials for the system.	WDA ConTA StrA SOCA WCA	The WDA phase of CWA can be used to define the requirements of training simulators. In addition, the WDA allows training to be developed that focuses on satisfying the functional purposes of the work domain, rather than training that describes specific sequences of behaviours. Similarly, the ConTA, StrA and SOCA and WCA phases can also guide training development

Table 10.1   *Concluded*

Performance Assurance		
• Ensure that the system starts working correctly and continues to function as intended. • Determine new system requirements resulting from capabilities and deficiencies of the operational system.	WDA ConTA StrA SOCA WCA	CWA provides a model to investigate if the necessary control tasks, strategies, coordination structures, and operator's cognitive capabilities are available for the system to function properly. A WDA can be useful in showing that a particular system is no longer competitive in meeting the functional requirements of the work domain and is ready to be replaced. In addition, ConTA may be useful for revealing shortfalls in the relevance or effectiveness with which control tasks are carried out. The evaluation technique proposed in chapter 5 provides a basis for evaluating new capabilities.
Problem Investigation • Investigate accidents and incidents and determine how to prevent their recurrence.	WDA ConTA StrA SOCA WCA	The results from the various phases of CWA can shed light on the root causes of accidents and incidents in terms of whether the system failed to support the operators adequately.

Closing Remarks

When we started this book, filled with excitement and enthusiasm for a new challenge, we set out on a quest seeking the 'Holy Grail' that is the explicit link between analysis and design. At the end of the book, it is clear that this quest is not complete, like the Holy Grail, its mere existence is contentious. As Chapter 3 discussed, design is, without doubt, an art form. Its reliance on the designer's creativity means that it will never be a science. It is almost impossible to conceive that a process or technique will ever be developed, that makes a completely explicit link between analysis and design. Those with a 'designer's mind' will always make extraordinary links from problems to solutions, that can only be described by onlookers as 'black art'; truly inspired designers will continue to break established 'rules' to produce stunningly great designs. Although no amount of tuition, no methodology, approach or framework can turn a person into a designer, what can be instilled is an appreciation of what good design is. We hope that this book has proved beyond doubt that through its process of describing the system constraints, CWA can provide a unique framework for describing the system in a common language to analyse, design and evaluate the system. We feel certain that, as the CWA framework is developed further, and applied to more domains, this gap will continue to shrink.

Appendix
Can it be Taught?

Introduction

The success of any technique, framework or methodology is inextricably linked to its ability to be assimilated by those who intend to apply it. The success is also linked to its ability to be understood by those with an interest in the findings of third party analyses. This leads to the questions, 'can CWA be taught?' and 'how effective can the tuition be within a limited time period?' During the early years of the framework's development, knowledge was passed on from master to apprentice, this was mainly due to lack of exposure to the framework. Its teaching in a wider classroom setting to students of different backgrounds is something that has developed, far more, in recent years. Using an artefact based example of a personal diary, this appendix will address how the technique can be taught to undergraduate students and suggest ways of enhancing the learning process.

The Study

To investigate the ability of novices to grasp and apply the CWA technique a small study was conducted in which final year students of Design at Brunel University were taught the process of Work Domain Analysis (WDA). The formal training consisted of a one hour lecture followed by a one hour seminar session; the students were divided into two groups of approximately twelve and taught the same material; 'Group A' the first week and 'Group B' the second week. The students were introduced to the framework in its entirety; however, due to time constraints, a decision was made not to teach the framework in detail beyond the first phase of analysis; care was taken to explain to the students the limitations of not extending the approach beyond the first phase.

Design

The study was devised to determine the student's ability to learn how to construct an Abstraction Hierarchy (AH) based on a one-hour lecture followed by a one-hour seminar. The students' performance was evaluated using the section of their end of year course work related to the CWA framework.

Participants

Twenty-three participants took part in the study, they comprised of 19 males and 4 females. Participants were all in their final year of Design at Brunel University (Uxbridge, UK) the participants all studied an elective module titled 'Cognitive Ergonomics'. The module accounted for 20 out of the year's 120 credits. When considered against the three year course (two of which count towards the final grade) the module accounts for just over 8 per cent of the student's final degree classification.

Experimental Methods

The participants were taught by the author of this book for one hour in a lecture style. The lecture, delivered with the aid of PowerPoint, covered a broad description of CWA before focusing in on the WDA stage. The tuition involved description of the Abstraction Hierarchy along with how to construct the WDA product from scratch. The lecture agenda was as follows:

- Overview outlining how CWA was different from other normative or descriptive approaches.
- Introduction and brief explanation of the five phases of CWA.
- Detailed introduction to Work Domain Analysis (WDA).
- Procedure for conducting a WDA of an analogue wrist watch.
- Advantages of the approach.
- Disadvantages of the approach.
- Summary.
- Further Reading.
- Questions from the students.

Following the lecture the participants moved rooms to a seminar setting, the participants were asked to construct their own AH for a given product; a paper-based diary and an electronic organiser. Following the study, the data was collected in and the students were provided with a short paragraph giving feedback on their attempts.

Equipment

Participants were provided with paper copies of the presented PowerPoint presentation along with A3 blank templates containing the Abstraction Hierarchy levels (see example Figure A.1). Participants were also provided with a ten page summary of CWA and WDA, the summary document contained hyperlinks to a number of internet based resources including a link to the Defence Science and

Technology Organisation (DSTO) CSWA website (http://www.dsto.defence.gov.au/research/page/3736/) and the Center of Human Information Interaction CWA portal (http://projects.ischool.washington.edu/chii/portal/index.html).

Procedure

The participants were required to complete an analysis of the two products; the paper-based diary and the electronic organiser. The participants were familiar with these products, as they had analysed them with other human factors approaches in the past as part of the course. The remit of the analysis was deliberately unconstrained with the participants free to define the overall functional purpose of the system to whatever they deemed appropriate. The task of setting the system boundaries was also left to the students. For this reason, the products generated were not expected to be identical.

The deliverable for the students was a section in their module course work on CWA and the analyses that they conducted. The section was marked by the author of this book and then independently moderated by the course tutor.

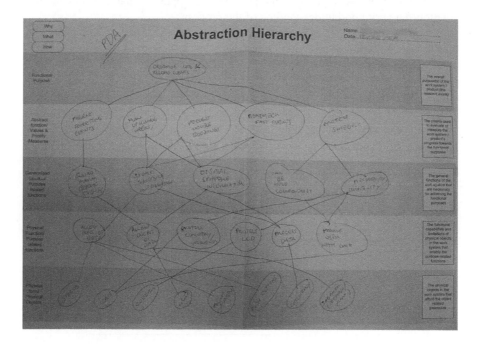

Figure A.1   Example of template provided to students

Results

The results were encouraging with the vast majority of the students producing correct AHs and demonstrating their understanding through their commentary of the products. Out of the 23 students, one failed to submit any kind of report and one failed to submit the AH in the appendix of the document. The resulting discussion therefore covers the 21 submitted complete reports.

Table A.1 breaks down the students results into three categories. Those with minimal errors in the construction of the AH, indicating an almost complete understanding of the technique; those with minor errors, indicating that the basic concept had been grasped but further explanation was needed to clarify the technique; and finally, those who showed major errors, indicating that there were fundamental concerns with the students understanding that would require significant explanation.

Figure A.2 shows an example of an Abstraction Hierarchy that contains only minimal errors, here the student has really grasped the concept of the AH process, they have appropriate nodes in the correct levels and have demonstrated a good understanding of how to apply the means-ends links. If the hierarchy is heavily scrutinised, there are some areas that are slightly awkward and not completely correct; however, as an early attempt it seems apparent that the student has developed a very good understanding of the process.

Figure A.3 shows an example with what has been termed as 'minor errors', the student has, with the exception of a few nodes, assigned nodes to their correct level. There seems to be some ambiguity in the description of the values and priority measures (abstract functions) for the 'record events' and the 'recall events' nodes, these nodes would fit more comfortably in the purpose related functions level (generalised functions). The student's understanding of the means-ends links also requires additional attention.

Table A.1   **Results of students error rates in completing an Abstraction Hierarchy**

Level of errors	Number of students
Minimal	5
Minor	15
Major	1
Total	**21**

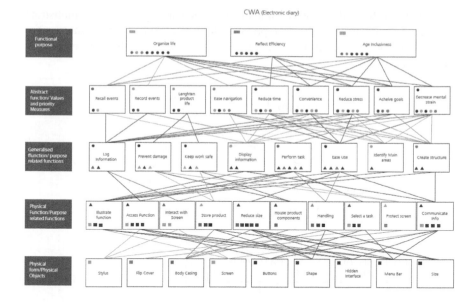

Figure A.2   An example of an Abstraction Hierarchy with minimal errors

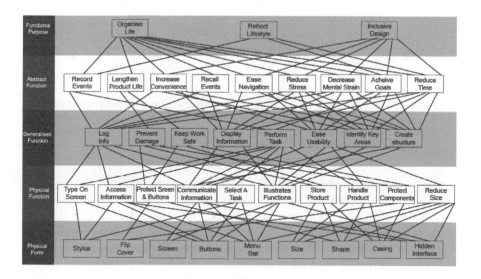

Figure A.3   An example of an Abstraction Hierarchy with minor errors

Figure A.4 shows the only example of a student that was unable to demonstrate a good understanding of the AH process, here the student has made a very limited attempt to develop the model or test the links using the how-what-why relationship.

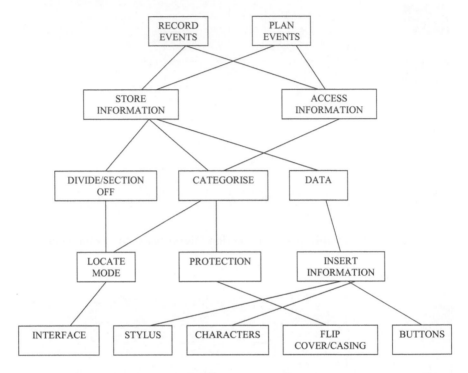

Figure A.4   An example of an Abstraction Hierarchy with major errors

Summary

In this exploratory study, the quality of the students' course-work was used to investigate the students' ability to learn the basics of WDA and construct an Abstraction Hierarchy for a familiar product. The results showed that the vast majority of the students were able to grasp the basic concepts of the approach in what was a very limited period. However, none of the students were able to create a flawless Abstraction Hierarchy; this indicates that further guidance and tuition beyond the two hours provided (or different lecture content) is required. With the exception of one, the students were capable of selecting appropriate analysis boundaries, using their chosen functional purpose(s) to reflect this. The students all demonstrated the ability to classify the components at the physical objects level correctly. The results suggest that the concept of representing the same domain at a number of levels of abstraction was well understood. By incorrectly categorising nodes, a significant number of students demonstrated an incomplete understanding of the levels of abstraction, this confusion was generally focused around the middle three levels of the hierarchy (Values and priority measures, Purpose related processes, Object related processes). The connection of the nodes with means-ends links was perhaps the least well conducted part of the process. From analysis of the submitted abstraction hierarchies, it is postulated that many of the students did not adequately validate the links using the why-what-how triad. This failure to complete this validation adequately could be at least partly redressed by a greater emphasis on the importance of this process in the lecture and seminar.

The task of selecting the boundary and scope for the analysis is considered a fundamentally important part of the analysis process. For this reason, a conscious decision was made not to prescribe this to the students. As a result, it is difficult to make comment on the reliability and repeatability of the technique. Because of different analysis boundaries, each of the submitted analysis proved to be unique. Reliability and validity is often sited as measure for evaluating different human factors methods (Stanton et al, 2005). Stanton et al (2005) suggest that the reliability and validity of the CWA framework is difficult to assess; Militello and Hutton (2000) suggest that there are no well-established metrics that exist to establish the reliability and validity of cognitive task analysis methods. With such a formative approach, however, it is contended that reliability is far less important than the flexibility to apply an appropriate system and analysis boundary.

Returning to the questions posed in the introduction to this appendix, 'can CWA be taught?' and 'how effective can the tuition be within a limited time period?' The results indicate that the approach, at a basic level, can be taught. It is also clear that the basics can be communicated in a limited period. The results indicate that to ensure a full understanding a more comprehensive training package would be required, however, it is contended that with additional self-study the WDA approach is teachable with only a few hours' tuition.

Bibliography

Ahlstrom, U. (2005), Work domain analysis for air traffic controller weather displays, *Journal of Safety Research*, 36, 159–169.

Ainsworth, L.K. and Marshall, E. (1998), Issues of quality and practicability in task analysis: preliminary results from two surveys. *Ergonomics* 41, 11, 1607–1617.

Annett, J. (1996), Recent developments in hierarchical task analysis, In: Roberson, S.A. (ed.), *Contemporary Ergonomics*, London: Taylor and Francis, 263–268.

Annett, J. (2004), Hierarchical task analysis, In: Diaper, D., Stanton, N.A. (eds), *The Handbook of Task Analysis for Human-Computer Interaction*, Mahwah, NJ: Lawrence Erlbaum Associates, 67–82.

Annett, J., Duncan, K.D., Stammers, R.B. and Gray, M.J. (1971), Task analysis, *Department of Employment Training Information Paper 6*, London: HMSO.

Apple, (2007a), 100 Million iPods Sold, Press release (online). Available http://www.apple.com/pr/library/2007/04/09ipod.html, accessed on the 15th of May 2007.

Apple, (2007b), Apple Unveils Higher Quality DRM-Free Music on the iTunes Store, Press release (online). Available http://www.apple.com/pr/library/2007/04/02itunes.html, accessed on the 15th of May 2007.

Bainbridge, L. (1987), Ironies of automation, In *New Technology and Human Error* (eds), J. Rasmussen, K. Duncan, and J. Leplat, New York: Wiley.

Bar Yam, Y. (1997), *Dynamics of Complex Systems*, Jackson, TN: Perseus.

Barnett, B.J. and Wickens, C.D. (1988), Display proximity in multi cue information: the benefits of boxes, *Human Factors*, 30, 15–24.

Bertalanffy, L.V. (1950), The theory of open systems in physics and biology, *Science*, 111, 23–29.

Beuscart, J.M. (2005), Napster users between community and clientele: The formation and regulation of a sociotechnical group, *Sociologie du Travail*, 47, S1–S16.

Bisantz, A.M., Roth, E., Brickman, B., Gosbee, L.L., Hettinger, L. and McKinney, J. (2003), Integrating cognitive analyses in a large-scale system design process, *International Journal of Human-Computer Studies*, 58, 177–206.

Bonaceto, C., and Burns, K. (2003), Mapping the Mountains: A Survey of Cognitive Engineering Methods and Uses. Available at: http://mentalmodels.mitre.org/cog_eng/ce_sys_eng_challenges.htm, accessed on the 14th of August 2008.

Brehmer, B. (2007), Understanding the functions of C2 is the key to progress, *The International C2 Journal*, 1, 1, 211–232.

British Defence Film Library (BDFL) (2001), *Armoured Battlegroup in the Quick Attack # C003/00*, Movie.

Bruseberg, A. (2005), *Review of Cognitive Work Analysis for Further CTA Tool Development*, HFI DTC WP 2.3.

Bruseberg, A. and Lintern, G. (2007), Human factors integration for MODAF: needs and solution approaches, *Seventeenth Annual International Symposium of the International Council on Systems Engineering (INCOSE)*, San Diego, CA, USA.

Builder, C.H., Bankes, S.C. and Nordin, R. (1999), *Command Concepts: A Theory Derived from the Practice of Command and Control*, Santa Monica, CA: Rand.

Burns, C.M., Bisantz, A.M. and Roth, E.M. (2004), Lessons from a comparison of work domain models: representational choices and their implications, *Human Factors*, 46, 4, 711–727.

Burns, C.M., Bryant, D.J. and Chalmers, B.A. (2000), A work domain model to support shipboard command and control, *In International Conference on Systems, Man, and Cybernetics*, IEEE, Nashville, TN, USA, 2228–2233.

Burns, C.M. and Hajdukiewicz, J.R. (2004), *Ecological Interface Design*, Boca Raton, FL: CRC Press.

Card, S.K., Mackinlay, J.D. and Shneiderman, B. (1999), *Readings in Information Visualization: Using Vision to Think*, San Francisco: Morgan Kaufmann Publishers.

Carroll, J.M., Kellogg, W.A. and Rosson, M.B. (1991): The task-artifact cycle, In: Carroll, J.M., *Designing Interaction: Psychology at the Human-Computer Interface*, Cambridge University Press.

CAST (2007), *The Combat Estimate, Combined Armed Staff Trainer Guide*, Warminster: Land Warfare Centre.

Chernoff, H. (1973), The use of faces to represent points in K-dimensional space graphically, *Journal of the American Statistical Association*, 68, 342, 361–368.

Cherns, A.B. (1987), 'Principles of Sociotechnical Design Revisited', *Human Relations*, 40, 3, 153–162.

Chin, M., Sanderson, P. and Watson, M. (1999), Cognitive Work Analysis of the command and control work domain, *Proceedings of the 1999 Command and Control Research and Technology Symposium*, Newport: RI, United States Naval War College.

Crone, D.J., Sanderson, P.M. and Naikar, N. (2003), Using Cognitive Work Analysis to develop a capability for the evaluation of future systems, *Proceedings of the 47th Annual Meeting Human Factors and Ergonomics Society*, Denver, CO, 1938–1942.

Crone, D., Sanderson, P., Naikar, N. and Parker, S. (2007), Selecting sensitive measures of performance in complex multivariable environments, *Proceedings of the 2007 Simulation Technology Conference (SimTecT 2007)*, Brisbane, Australia, 4–7.

Cummings, M.L. and Guerlain, S. (2003), The tactical Tomahawk conundrum: Designing decision support systems for revolutionary domains, *IEEE Systems, Man, and Cybernetics Society Conference*, Washington DC, October 2003.

Cummings, M.L. and Guerlain, S. (in review), The Decision-ladder as an automation planning tool, *Cognition, Technology, and Work.*

Darses F. (2001), Providing practitioners with techniques for cognitive work analysis, *Theoretical Issues in Ergonomics Science,* 2:3, 268–277.

De Carvalho, P.V.R. (2006), Ergonomic field studies in a nuclear power plant control room, *Progress in Nuclear Energy,* 48, 51–69.

Dekker, A.H. (2003), Using agent-based modelling to study organisational performance and cultural differences, *Proceedings of the MODSIM 2003 International Congress on Modelling and Simulation,* Townsville, Queensland, 1793–1798. Available at www.mssanz.org.au/modsim03/Media/Articles/Vol 4 Articles/1793–1798.pdf.

EBO Prototyping Team (2005), *Effects-Based Approach to Multinational Operations; Concept of Operations (CONOPS) with Implementing Procedures,* Version 0.90.

Embrey, D.E. (1986), SHERPA: A systematic human error reduction and prediction approach, *Paper presented at the International Meeting on Advances in Nuclear Power Systems,* Knoxville, Tennessee, USA.

Fidel, R. and Pejtersen, A.M. (2005), Cognitive Work Analysis, In K.E. Fisher, S. Erdelez, E.F. McKechnie (eds), *Theories of Information Behavior: A Researcher's Guide,* Medford, NJ: Information Today.

Gorman, J.C., Cooke, N.J. and Winner, J.L., (2006), Measuring team situation awareness in decentralized command and control environments, *Ergonomics,* 49, 1312–1325.

Grand, S. (2000), *Creation: Life and How to Make It,* London: Orion.

Grether, W.F. (1949), The design of long-scale indicators for speed and accuracy of quantitative reading, *Journal of applied Psychology,* 33, 363–372.

Hajdukiewicz J.R. (1998), *Development of a Structured Approach for Patient Monitoring in the Operating Room,* Masters book, University of Toronto.

Hajdukiewicz, J.R., Burns, C.M., Vicente, K.J. and Eggleston, R.G., (1999), Work domain analysis for intentional systems, *Proceedings of the Human Factors and Ergonomics Society 43rd Annual Meeting.* 333–337.

Hajdukiewicz, J.R. and Vicente, K.J. (2002), Designing for adaptation to novelty and change: The role of functional information and emergent features, *Human Factors,* 44, 592–610.

Hajdukiewicz, J.R. and Vicente, K.J. (2004), A theoretical note on the relationship between work domain analysis and task analysis, *Theoretical Issues in Ergonomics Science,* 5, 6, 527–538.

Higgins, P.G. (1998), Extending Cognitive Work Analysis to Manufacturing Scheduling, In P. Calder and B. Thomas (eds), *Proceedings 1998 Australian Computer Human Interaction Conference,* OzCHI'98, November 30–December 4, Adelaide, IEEE, 236–243.

Hollnagel, E. (1992), Coping, coupling and control: the modelling of muddling through, *In Proceedings of 2nd Interdisciplinary Workshop on Mental Models,* March 25–26, Cambridge, UK, 61–73.

Hollnagel, E. (1995), Automation, coping, and control, S. Kondo (ed.), *Post HCI '95 Seminar on Human-Machine Interface in Process Control – Interface for Co-understanding and Cooperation*, July 17–18, Hieizan, Japan.

Houghton, R.J., Baber, C., McMaster, R., Stanton, N.A., Salmon; P.M., Stewart, R. and Walker, G.H. (2006), Command and control in emergency services operations: A social network analysis, *Ergonomics*, 49, 1204–1225.

JDCC (Joint Doctrine and Concepts Centre) (2005), *The UK Military Effects-Based Approach*, Joint Discussion Note 1/05.

JDCC (Joint Doctrine and Concepts Centre) (2006), *The Comprehensive Approach*, Joint Discussion Note 4/05.

Jamieson, G.A. and Vicente, K.J., (2001), Ecological interface design for petrochemical applications: Supporting operator adaptation, continuous learning, and distributed, collaborative work, *Computers and Chemical Engineering*, 25, 1055–1074.

Jansson, A., Olsson, E. and Erlandsson, M. (2006), Bridging the gap between analysis and design: Improving existing driver interfaces with tools from the framework of cognitive work analysis, *Cognition, Technology and Work*, 8, 1, pp. 41–49.

Jenkins, D.P., Farmilo, A., Stanton, N. A., Whitworth, I., Salmon, P.M., Hone, G., Bessell, K. and Walker, G.H. (2007a), *The CWA Tool V0.95*, HFI DTC, Yeovil, Somerset, UK.

Jenkins, D.P., Stanton, N.A., Salmon, P.M, Walker, G.H., Young, M.S. Farmilo, A., Whitworth, I. and Hone, G. (2007b), The development of a cognitive work analysis tool, In D. Harris (ed.), *Engin. Psychol. and Cog. Ergonomics*, HCII 2007, LNAI 4562, 504–511.

Jenkins, D.P., Stanton, N.A., Walker, G.H. and Salmon, P.M. (2007c), Cognitive work analysis of a sensor to effecter system: Implications for network structures, In D. de Waard, G.R.J. Hockey, P. Nickel, and K.A. Brookhuis (eds) (2007), *Human Factors Issues in Complex System Performance* (73–84), Maastricht, the Netherlands: Shaker Publishing.

Jenkins, D.P., Stanton, N.A., Walker, G.H., Salmon, P.M. and Young, M.S. (2008a), Using cognitive work analysis to explore activity allocation within military domains, *Ergonomics*, 51, 6, 789–815.

Jenkins, D.P., Stanton, N.A., Walker, G.H., Salmon, P.M. and Young, M.S. (2008b), Applying cognitive work analysis to the design of rapidly reconfigurable interfaces in complex networks, *Theoretical Issues in Ergonomics Science*, 9, 4, 273–295.

Jenkins, D.P., Stanton, N.A., Walker, G.H. and Young, M.S. (2007d), 'A new approach to designing lateral collision warning systems', *Int. J. Vehicle Design*, 45, 3, 379–396.

Jobs, S. (2007a), Macworld San Francisco 2007, Keynote Address (online). Available http://www.apple.com/quicktime/qtv/mwsf07/, accessed on the 14th of December 2007.

Jobs, S. (2007b), Thoughts on Music (online). Available http://www.apple.com/hotnews/thoughtsonmusic/, accessed on the 14th of December 2007.

Kasik, D.J. (2002), Evaluating graphics displays for complex 3D models, *IEEE Computer Graphics and Applications*, 22, 3, 56–64.

Kilgore, R. and St-Cyr, O. (2006), SRK Inventory: A tool for structuring and capturing a worker companies analysis, *Proceedings of the Human Factors and Ergonomics Society 50th Annual Meeting 2006*, 506–509.

Kirwan, B. and Ainsworth, L.K. (eds) (1992), *A Guide to Task Analysis*, London, UK: Taylor and Francis.

Klein, G.A., Wolf, S., Militello, L. and Zsambok, C. (1995), Characteristics of skilled option generation in chess, *Organizational Behavior and Human Decision Process*, 62, 1, 63–69.

Lamoureux, T.M., Rehak, L.A., Bos, J.C. and Chalmers, B. (2005), Control task analysis in applied settings, *Proceedings of the Human Factors and Ergonomics Society 50th Annual Meeting 2006*, 391–395.

Lintern, G. (2005), Work Domain Analysis: Tutorial (online). Available http://cognitivesystemsdesign.net/Tutorials/WDA%20Tutorial.pdf, accessed on the 14th of August 2008.

Lintern, G. (2006), A functional workspace for military analysis of insurgent operations, *International Journal of Industrial Ergonomics* 36, 409–422.

Lintern, G., Cone, S., Schenaker, M., Ehlert, J. and Hughes, T. (2004), *Asymmetric Adversary Analysis for Intelligent Preparation of the Battlespace (A3-IPB)*, United States Air Force Research Department Report.

Lintern, G. and Naikar, N. (2000), The use of work domain analysis for the design of training teams, *Proceedings of the joint 14th Triennial Congress of the International Ergonomics Association/44th Annual Meeting of the Human Factors and Ergonomics Society (HFES/IEA 2000)*, San Diego, CA.

Memisevic, R., Sanderson, P., Choudhury, S. and Wong, W. (2005), Work domain analysis and ecological interface design for hydropower system monitoring and control, *In Proceeding of the IEEE Conference on Systems, Man, and Cybernetics (IEEE-SMC2005)*, Hawaii, USA, 3580–3587.

Militello, L. and Hutton, R. (2000), Applied cognitive task analysis (ACTA): a practitioner's toolkit for understanding cognitive task demands, In Annett and Stanton (2000), 90–113.

Miller, C.A. (2004), A work domain analysis framework for modelling intensive care unit patients, *Cognition, Technology and Work*, 6, 4, 207–222.

Miller, C.A. and Vicente K.J. (2001), Comparison of Display Requirements Generated via Hierarchical Task and Abstraction-Decomposition Space Analysis Techniques, *International Journal of Cognitive Ergonomics*, 5, 3, 335–355.

Mohnkern, K. (1997), Beyond the Interface Metaphor (online). Available http://www.cs.cmu.edu/afs/cs.cmu.edu/user/kem/www/vid/vid9704.html, accessed on the 14th of August 2008.

Moore, G.E. (1965), Cramming more components onto integrated circuits, *Electronics*, 38, (8).

Muckler, F.A. and Seven, S.A. (1992), Selecting performance measures: 'Objective' versus 'subjective' measurement, *Human Factors*, 34, 4, 441–455.

Munson, R.C. and Horst, R.L. (1986), Evidence for global processing of complex visual display, *In Proceedings of the Human Factors Society 30th Meeting*, Santa Monica, CA: Human Factors Society, 776–780.

Naikar, N. (2006b), Beyond interface design: Further applications of cognitive work analysis, *International Journal of Industrial Ergonomics*, 36, 423–438.

Naikar, N. (2006c), An examination of the key concepts of the five phases of cognitive work analysis with examples from a familiar system, *Proceedings of the Human Factors and Ergonomics Society 50th Annual Meeting*, 447–451.

Naikar, N. (in press), Modelling activity in network-centric operations with cognitive work analysis: Work situations, work functions, decisions, and strategies. In B. Bolia (ed.), *Supporting Decision Effectiveness in Network-Centric Operations*, Wright Patterson Air Force Base, Dayton, OH.

Naikar, N., Hopcroft, R. and Moylan, A. (2005), *Work Domain Analysis: Theoretical Concepts and Methodology*, DSTO-TR-1665.

Naikar, N. and Lintern, G. (2002), A review of 'Cognitive work analysis: Towards safe, productive, and healthy computer-based work' by Kim J. Vicente, *The International Journal of Aviation Psychology*, 12, 4, 391–400.

Naikar, N., Moylan, A. and Pearce, B. (2006), Analysing activity in complex systems with cognitive work analysis: Concepts, guidelines, and case study for control task analysis, *Theoretical Issues in Ergonomics Science*, 7, 4, 371–394.

Naikar, N. and Pearce, B. (2003), Analysing activity for future systems, *Proceedings of the 47th Annual Meeting of the Human Factors and Ergonomics Society*, Santa Monica, CA: Human Factors and Ergonomics Society, 1928–1932.

Naikar, N., Pearce, B., Drumm, D. and Sanderson, P.M. (2003), Technique for designing teams for first-of-a-kind complex systems with cognitive work analysis: Case study, *Human Factors*, 45, 2, 202–217.

Naikar, N. and Sanderson, P.M. (1999), Work domain analysis for training-system definition, *International Journal of Aviation Psychology*, 9, 271–290.

Naikar, N. and Sanderson, P.M. (2001), Evaluating design proposals for complex systems with work domain analysis, *Human Factors*, 43, 529–542.

Naikar, N. and Saunders, A. (2003), Crossing the boundaries of safe operation: A technical training approach to error management, *Cognition Technology and Work*, 5, 171–180.

NATO Naval Armaments group IEG/5-SG/8 Ninth Meeting. (1989), An evaluation of three tactical symbol set options, *Working Paper Three Evaluation*, TNO institute for perception 26–30 June, Soesterberg, Netherlands.

Norman, D. (1990), The problem of automation: Inappropriate feedback and interaction, not over-automation, *Philosophical Transactions of the Royal Society of London*, B, 1990.

Norman, D. (2002), *The Design of Everyday Things*, New York: Basic Books.

Norman, D. (2004), *Emotional Design: Why We Love (or Hate) Everyday Things*, New York: Basic Books.

Norman, D. (2007), *The Design of Future Things*, New York: Basic Books.

Olsson, G. and Lee, P.L. (1994), Effective interfaces for process operators, *The Journal of Process Control*, 4, 99–107.

Rasmussen, J. (1974), *The Human Data Processor as a System Component: Bits and Pieces of a Model (Report No. Risø-M-1722)*, Roskilde, Denmark: Danish Atomic Energy Commission.

Rasmussen, J. (1976), Outlines of a hybrid model of the process plant operator, *Monitoring Behavior and Supervisory Control*, G. Johannsen, Plenum.

Rasmussen, J. (1983), Skills, rules, knowledge; signals, signs, and symbols, and other distinctions in human performance models, *IEEE Transactions on Systems, Man and Cybernetics*, 13, 257–266.

Rasmussen, J. (1985), The role of hierarchical knowledge representation in decision making and system management, *IEEE Transactions on Systems, Man and Cybernetics*, 15, 234–243.

Rasmussen, J. (1986), *Information Processing and Human-machine Interaction: An Approach to Cognitive Engineering*, New York: North-Holland (online). Available http://www.ischool.washington.edu/chii/portal/literature.html, accessed on the 14th of August 2008.

Rasmussen, J. (1990), Mental models and the control of action in complex environments, In D. Ackermann and M.J. Tauber (eds), *Mental Models and Human Computer Interaction 1*, Amsterdam: North-Holland, 41–69.

Rasmussen, J. (1999), *Forward to Cognitive Work Analysis: Toward Safe, Productive, and Healthy Computer-based Work*, pp xi, Mahwah, New Jersey: Lawrence Erlbaum Associates.

Rasmussen, J., Pejtersen, A. and Goodstein, L.P. (1994), *Cognitive Systems Engineering*, New York: Wiley.

Rasmussen, J., Pejtersen, A.M. and Schmidt, K. (1990), *Taxonomy for Cognitive Work Analysis*, Risø National Laboratory Risø-M-2871.

Redström, J. (2006), Towards user design? On the shift from object to user as the subject of design, *In Design Studies*, 27, 2, 123–139.

Rehak, L.A., Lamoureux, T.M. and Bos, J.C. (2006), Communication, coordination, and integration of Cognitive Work Analysis outputs, *Proceedings of the Human Factors and Ergonomics Society 50th Annual Meeting*, 391–395.

Rittel, H. and Webber, M. (1973), Dilemmas in a General Theory of Planning, *Policy Sciences*, 4, 155–169 (Reprinted in N. Cross (ed.), Developments in Design Methodology, Chichester: J. Wiley and Sons, 1984, 135–144.)

Saffer, D. (2007), *Designing for Interaction: Creating Smart Applications and Clever Devices*, CA: New Riders in association with AIGA Design Press.

Salmon, P.M., Stanton, N.A., Regan, M., Lenne, M. and Young, K., (2007), Work domain analysis and road transport: Implications for vehicle design, *International Journal of Vehicle Design*, 45, 3, 426–448.

Salmon, P., Stanton, N., Walker, G. and Green, D. (2004), Future battlefield visualisation: Investigating data representation in a novel C4i system, In V. Puri, D. Filippidis, P. Retter and J. Kelly (eds), *Weapons, Webs and Warfighters. Proceedings of the Land Warfare Conference*, DSTO, Melbourne.

Sanderson, P.M. (2003), Cognitive work analysis across the system life-cycle: Achievements, challenges, and prospects in aviation, In P. Pfister and G. Edkins (eds), *Aviation Resource Management*, 3, Aldershot: Ashgate.

Sharp, T.D. and Helmicki A.J. (1998), The application of the ecological interface design approach to neonatal intensive care medicine, *In Proceedings of the Human Factors and Ergonomics Society 42nd Annual Meeting Santa Monica*, CA: HFES, 350–354.

Shepherd, A. (1985), Hierarchical task analysis and training decisions, *Programmed Learning and Educational Technology*, 22, 162–176.

Shneiderman, B. (1983), Direct manipulation: A step beyond programming languages, *IEEE Computer*, 16, 8, 57–69.

Shneiderman, B. (1998), *Designing the User Interface*, third ed. Reading, Mass: Addison-Wesley.

Sitter, L.U., Hertog, J.F. and Dankbaar, B. (1997), From complex organizations with simple jobs to simple organizations with complex jobs, *Human Relations*, 50, 5, 497–536.

Smith, E. (2003), *Effects Based Operations, Applying Network Centric Warfare in Peace, Crisis*.

Stanton, N.A. (2006), Hierarchical task analysis: Developments, applications, and extensions, *Applied Ergonomics*, 37, 55–79.

Stanton, N.A., Harris, D., Salmon, P., Demagalski, J.M., Marshall, A., Young, M.S., Dekker, S.W.A. and Waldmann, T. (2006), Predicting design induced pilot error using HET (Human Error Template) – A new formal human error identification method for flight decks, *Aeronautical Journal*, 110, 107–115.

Stanton, N.A., Salmon, P.M., Walker, G.H., Baber, C. and Jenkins, D. (2005), *Human Factors Methods: A Practical Guide for Engineering and Design*, Aldershot: Ashgate.

Stanton, N.A., Stewart, R., Harris, D., Houghton, R.J., Baber, C., McMaster, R., Salmon, P.M., Hoyle, G. Walker G.H., Young, M.S., Linsell, M., Dymott, R. and Green, D., (2006), Distributed situation awareness in dynamic systems: theoretical development and application of an ergonomics methodology, *Ergonomics*, 49,1288–1311.

Staples, L.J. (1993), The task analysis process for a new reactor, *In: Proceedings of the Human Factors and Ergonomic Society 37th Annual Meeting-Designing for Diversity*, Seattle, WA, October 11–15, The Human Factors and Ergonomic Society, Santa Monica CA 1024–1028.

Storr, J. (2005), A critique of effects-based thinking, RUSI Journal.

T3, (2006), Review of Apple iPod 80GB (online). Available http://www.t3.co.uk/ reviews/entertainment/mp3_player/apple_ipod_80gb, accessed on the 14th of August 2008.

The Gadget Show, (2007), Best Buys... MP3 players - 30/3/07 (online). Available http://gadgetshow.five.tv/jsp/5gsmain.jsp?lnk=501&featureid=352&descripti on=MP3%20players, accessed on the 14th of August 2008.

Trist, E. and Bamforth, K. (1951), Some social and psychological consequences of the longwall method of coal getting, *Human Relations*, 4, 3–38.

Troy, M. and Möller, T. (2004), Human factors in visualization research, *IEEE Transactions on Visualisation and Computer Graphics*, 10, (1).

Tufte, E.R. (1983), *The Visual Display of Quantitative Information*, Cheshire, CT: Graphics Press.

Tufte, E.R. (1990), *Envisioning Information*, Cheshire, CT: Graphics Press.

Tufte, E.R. (1997), *Visual Explanations; Images and Quantities, Evidence and Narrative*, Cheshire, CT: Graphics Press.

Vicente, K.J. (1997), Interface design: Is it always a good idea to design an interface to match the operator's mental model?, *Ergonomics Australia On-Line*, 11, 2.

Vicente, K.J. (1999), *Cognitive Work Analysis: Toward Safe, Productive, and Healthy Computer-based Work*, Mahwah, NJ: Lawrence Erlbaum Associates.

Vicente, K.J. (2002), Ecological interface design: Progress and challenges, *Human Factors*, 44, 62–78.

Walker, G.H., Gibson, H., Stanton, N.A., Baber, C., Salmon, P.M. and Green, D. (2006a), Event analysis of systemic teamwork (EAST): a novel integration of ergonomics methods to analyse C4i activity, *Ergonomics*, 49, 1345–1369.

Walker, G.H., Stanton, N.A., Jenkins, D., Salmon, P.M. (2006b), *Using an Electronic C4i System to Examine the Effects of Information Source and Decay*, Yeovil: Aerosystems.

Walker, G.H., Stanton, N.A., Young, M.S., Jenkins, D.P. and Salmon, P.M. (2007), *Review of the Knowledge Base: Sociotechnical Systems in NEC Design*, Technical Report.

Watson, M.O. and Sanderson, P.M. (2007), Designing for attention with sound: Challenges and extensions to ecological interface design, *Human Factors*, 49, 2, 331–346.

Ware, C. (2000), *Information Visualization: Perception for Design*, San Francisco: Morgan Kaufmann (Academic Press), 2000.

Wickens, C.D. (1992), *Engineering Psychology and Human Performance (2nd ed.)*, New York: Harper-Collins.

Woods, D.D. (1991), The cognitive engineering of problem representations. In G.R.S. Weir and J.L. Alty (eds), *Human-Computer Interaction and Complex Systems*, Academic Press ,169–188.

Woods, D.D. (1995), Toward a theoretical base for representation design in the computer medium: Ecological perception and aiding human cognition. J.M. Flach, P.A. Hancock, J. Caird, and K.J. Vicente (eds), *Global Perspectives on the Ecology of Human-Machine Systems*, Vol. 1, Resources for ecological psychology, pp. 157–188, Lawrence Erlbaum.

Wong, W.B.L., Sallis, P.J. and O'Hare, D. (1998), The ecological approach to interface design: Applying the abstraction hierarchy to intentional domains, *In*

The Eighth Australian Conference on Computer-Human Interaction OzCHI'98, P. Calder and B. Thomas, (eds), Adelaide, Australia: IEEE Computer Society Press, 1998, pp. 144–151.

Xu, W. (2007), Identifying Problems and Generating Recommendations for Enhancing Complex Systems: Applying the Abstraction Hierarchy Framework as an Analytical Tool, *Human Factors*, 49, 6, 975–994.

Yu, X., Lau, E., Vicente, K.J. and Carter, M.W. (2002), Toward theory-driven, quantitative performance measurement in ergonomics science: the abstraction hierarchy as a framework for data analysis, *Theoretic Issues in Ergonomic Science*, 3, 2, 124–142.

Zsambok, C.E. and Klein, G. (eds). (1997), *Naturalistic Decision Making*, Mahwah, NJ: Erlbaum.

Index

Other titles from Ashgate

Macrocognition in Teams:
Theories and Methodologies
Michael P. Letsky, Norman W. Warner, Stephen M. Fiore and
C.A.P. Smith
2008 • 436 pages
Hardback
978-0-7546-7325-5

Modelling Command and Control:
Event Analysis of Systemic Teamwork
Neville A. Stanton, Chris Baber and Don Harris
2008 • 274 pages
Hardback
978-0-7546-7027-8

Performance Under Stress
Peter A. Hancock and James L. Szalma
2007 • 406 pages
Hardback
978-0-7546-7059-9

ASHGATE